FLY ASH AMENDMENT AND PLANT GROWTH

FLY ASH AMENDMENT
AND
PLANT GROWTH

Dr. (Mrs.) SUMEDHA MEHTA MOHAN
Lecturer
Amity Institute of Biotechnology
Amity University (U.P.)

KALPAZ PUBLICATIONS
DELHI - 110052

Fly Ash Amendment and Plant Growth

Rs. 690

ISBN: 81-7835-457-8

Published in 2005 in India by
Kalpaz Publications
C-30, Satyawati Nagar
Delhi-110052
e-mail: kalpaz@hotmail.com

Laser Typeset at: Best Computers, Delhi
Printed at: Salasar Imaging System, Delhi

Dedicated to

My

Parents

My

Family

Contents

Preface

The thermal power projects ensured a sense of pride for people but now have become a threat for mankind. Most of the energy demands in developing countries like India are met by fossil fuels, coal being the major source. But besides providing energy it also provides urban air pollution, acid rain and potential global warming. Coal combustion also produces noxious gases like sulphur dioxide, nitrogen oxides and variety of residues like fly ash.

Fly ash constitutes the major portion of the total quantity of residues produced in power plants. The large amount of fly ash that is generated each year calls for a great deal of research to determine its feasibility for various potential uses.

Disposing off the huge amounts of fly ash is a major concern. Its use in manufacture of cement, concrete and bricks needs to be encouraged and requires research and developmental efforts with cost consideration. Changes in the soil characteristics and the cropping system due to addition of fly ash needs to be studied and investigated. It is desirable to establish vegetation on the landfill for stabilization against wind and water erosion thus providing shelter and habitat for wildlife and forming an aesthetically pleasing landscape.

Of all the possible uses of fly ash in a productive manner, the best use is as a fertilizer either through soil amendment or reclamation of fly ash dumps by cultivating plants. Efforts have been made in the present venture to modify the properties of agricultural soil by addition of fly ash, thus improving the soil fertility and crop yield. Identification of plants capable of doing well in presence of fly ash has become necessary for getting sustained yields under the changed scenario.

Considering the potential environmental impacts of fly ash, the present study aims at investigating the influence of fly ash generated from Badarpur Thermal Power Station (BTPS) on two economically and medicinally important trees viz. *Cordia dichotoma* Forst F. and *Morus alba* Linn. Different concentration of fly ash amendments in soils are used to ascertain the impact on both the species against the control-unamended soil.

Dr. (Mrs.) SUMEDHA MEHTA MOHAN

Acknowledgement

I bow in reverence to *Almighty God,* the omnipotent and omniscient, who has been the power behind my research. His mercy, compassion and blessings have guided me at each and every step that enabled me to complete this work.

I candidly feel that the words will hardly registered the message of indebtedness orginating from the core of my heart for ***Dr. Ashok K. Chauhan,*** Founder President, A.K.C. Group and Amity University for showing inmense confidence in the Faculty Members which enabled me to tide over my difficulties and obtain my objective. I express my sincere gratitude to ***Dr. Sunil Saran,*** Director, Amity Institute of Biotechnology and ***Prof. A.K. Srivastava,*** Dean, Amity Institue of Biotechnolgy, Amity University for lending constant support and for the benefit derived from their vast experiences. I am beholden to my seniors, colleagues and all the staff members of Amity Instiute of Biotechnology for their untiring support, love and affection.

With due respect and great pleasure. I express my heartfelt gratitude to ***Professor Mohammad Iqbal, Dr. Mahmooduzzafar, Dr. T.O. Siddiqui and Professor P.S. Srivastava,*** Hamdard University, Delhi for their scholarly suggestions and valuable guidance which enabled me to take this research work to its logical conclusion.

The morale boosting words, friendly gestures and wise advises rendered by the best of friends anyone can have, ***Malini Jacob,*** were a constant source of encouragement during my low periods.

My husband, ***Deep*** and my sons ***Vansh*** & ***Ishu*** both are the backbone of my research and publishing work. The understanding shared by my husband and the unparalleled smile and patience of my two sons have been and will always be the strength in my life. I thank all my in laws for their constant support and cooperation. Above all, I pay an honest and humble tribute to Papa, ***Late Mr. V.M. Mohan,*** a thorough gentleman for all this heavenly blessings. Though he is not around but he will always be my source of inspiration.

I feel honoured in thanking my relatives and family friends ***Duttas, Kapoorias, Mehtas, Sharmas, Sehgals, Mohans, Balis, Neelavaras, Sambhars*** and ***Qaziji*** for lending heartfelt support and good wishes through my tough and hard times.

Lastly but above all I bow with gratitude for my family, my constant support and source of confidence and encouragement. The warmth of the family given by my ***Papaji and Mummy,*** who always cheered me and lightened my mood in tiring hours will always be treasured by me. ***"God bless every child with such a loving family."***

SUMEDHA MEHTA MOHAN

1

Introduction to Fly Ash

Nature presents a synthesis of air, water, fire and soil. Man has been living in a balanced harmony with nature since time immemorial. His necessities, knowledge and values largely determine the changes which man has brought about in his environment. Although our environment has never been free from natural stresses such as droughts, frosts, excessive rainfall etc., but it never degraded as much as it does today. The emphasis has now shifted from the risks that the environment poses for our species to the risks that our species pose to our environment. Population explosion coupled with urbanization and industrialization has filled the environment with innumerable anthropogenic stresses, which include a variety of pollutants of air, water and soil. These man-made stresses influence not only the living beings but the non-living entities also. Environmental degradation is certainly a global phenomenon, but developing countries are the worst sufferers.

The application of thermal power projects was motivated by the need to create a better social development and a sense of pride for people, but unfortunately it has become a threat to life and environment. Over the last century, the consumption of sulphurous fuels such as coal and petroleum has raised manifold to meet the ever-increasing demand for energy and transport. The debate on environment and development has assumed a phenomenal importance at the global as well as national level. There is a virtual unanimity that economic development and sound environmental management complement each other given that on one side development itself contribute to environmental regeneration and rejuvenation, and that on the other, a healthy environment is a necessary condition for sustainable development.

Most of the energy demands in today's world is met by fossil fuel. 90% of the energy, we use, are provided by fossil fuels, coal being the major source. Besides providing energy, it also provides urban air pollution, acid rain, and potential global warming. Coal combustion also produces gases like sulphur dioxide, nitrogen, carbon dioxide etc. and a variety of residues like fly ash, bottom ash, flue gas, desulfurization waste, fluidized bed boiler waste and coal gasification waste.

Fly ash constitutes over 70% of the total quantity of residues produced in power plants. Because such a large amount of fly ash is generated each year, a great deal of research needs to be conducted to determine the feasibility of its utilization in agriculture, construction and other industries. Rough estimates of existing utilization are around 10% (1998) of the total fly ash generated as against 3-5% (1988).

From chemical viewpoint all naturally occurring trace elements enrich fly ash such as As, B, Ca, Mo, S, Se, Sr and V. Carbon (C) and Nitrogen (N) are usually present in small amounts, whereas levels of Potassium (K) and Phosphorous (P) are very high. Application of fly ash to soil alters the physical, chemical and biological properties of the soil. The major impacts of ash disposal are leaching of toxic substances from ash into soils and ground water, reduction in plant establishment and growth, increase of elements in vegetation growing on fly ash and transfer of elements through food chain.

The world is facing a real dilemma, because there is no alternative- either to find a choice to coal and other fossil fuels, or go ahead and burn all the fossil fuels resulting in a degraded planet. The major commercial electricity generation in the world is through fossil fuels. Total electricity generation in the world is around 2.6 million MW (1990) of which fossil fuel (thermal power) account for about 64%, hydropower 21% and nuclear power around 25%. The problem of pollution begins with the coal itself. Coal has reemerged as a sustainable energy source for the next decade due to ever increasing demand for energy, uncertainties associated with resources of petroleum and natural gas, inherent problems with nuclear power plants and current unfavourable economics of solar energy and biomass utilization.

By the mid of the twentieth century, coal was the world's dominant fuel (Griffith and Clarke, 1979) and it is still for power generation in India (Pathak *et al.*, 1989). Coal being the largest naturally occurring source of commercial energy, meets over 50% of the energy demand. The total gross reserves of coal are estimated at 85,075 million tonnes. Production of coal in India rose from a meagre, 30 million tonnes in 1950 to over 104 million tonnes in 1960. About 62% of the coal produced in our country is utilized for generation of electricity that account for 65% of power generation. The combustion of coal increased significantly due to super thermal power stations that came up and rose to > 200 million tones by 2000 AD (Singh and Yunus, 2000).

Indian coals are believed to be of drift origin and have a very high ash content (Ghose and Banerjee, 1995). The inorganic residues that remain after pulverized coal is burnt, are known as Coal Combustion Byproducts (CCBs). CCBs rapidly accumulate and cause enormous problem of disposal in the environment (El Mogazi *et al.*, 1988; Vallack and Chadwick, 1992; Carlson and Adriano, 1993). CCBs include a variety of residues including fly ash, bottom ash, flue gas, desulfurization waste, fluidized bed boiler waste and coal gasification ash (Singh and Yunus, 2000). Fly ash is the residue from coal combustion that constitutes over 70% of the total quantity of residues produced in power plants (Anonymous, 1997). It is estimated that nearly 2 tonnes of fly ash is produced every minute at a coal driven 200MW capacity generating unit (Kuchanwar and Matte, 1997).

Interaction between plants and fly ash amended soil are complex. Low concentration of fly ash in soil may enhance plant growth but higher concentrations are toxic. Fly ash deposits on leaves affect the stomatal functioning, transpiration and photosynthesis. Plants growing in fly ash affected areas accumulate elements in high concentration which are passed on to those feeding on the plants. The plants that are good Nitrogen fixers and tolerant to high salinity and Boron levels are generally the most successful early colonizers on fly ash landfills.

Fly ash is defined as the portion of ash produced in coal combustion That has a sufficiently small particle size to be carried away from the boiler in the flue gas. Being fine in particle size, it is a readily escape to the atmosphere along with flue gas and become a source of air pollution. In India, about 65 million tonnes of coal ash was produced during 1993 and is likely to exceed 140 million tonnes by the year 2020 AD (Anonymous, 1997). Fly ash is either collected in emission control devices, such as electrostatic precipitators and or mechanical filters, or released from the stack. Inspite of 99% efficiency of the collection, some very fine particles are emitted in the atmosphere (Page *et al.*, 1978). Fly ash collected by the emission control devices is disposed off in settling ponds and/or landfills. Landfills are the areas that receive fly ash directly from the power station (USE PA., 1988). Settling ponds receive ash in wet slurry form while in dry form it is released from the stack. The former may lead to ground surface water pollution primarily due to heavy metals present in the ash. The latter involves chances of ash being carried to far off distances due to wind and water action (Iqbal *et al.*, 2000).

Fly ash is a major contributor for soil pollution and alters the physical, chemical and biological properties of the soil. Properties of fly ash depend on the composition of the parent coal, conditions during coal combustion, efficiency of emission control devices, storage and handling of the byproducts and climate (van Hook, 1979; Adriano *et al.*, 1982). It is therefore difficult to generalize the composition of ashes or their behaviour in the environment, however certain characteristics are fairly uniform for most ashes (Table 1). About 60% of the fly ash samples have flowy consistency and 40% are of granular texture (Furr *et al.*, 1977). Fly ash consists of many minute glass like particles of 0.01 to 100mm size with specific gravities 2.1 to 2.6 (Davison *et al.*, 1974). The spherical, glassy and transparent appearance of fly ash indicates the presence of silicate minerals during coal combustion. Some spheres are filled with small amorphous particles and crystals (Hodgson and Holliday, 1966).

Coal is known to contain every natural occurring element, so it is not surprising that many trace elements are also found in fly ash (Page *et al.*, 1978; Klein *et al.*, 1975; Kaakien *et al.*, 1975). Among the elements generally enriched in ashes are As, B, Ca, Mo, S, Se and Sr (Page and Chang, 1979; El Mogazi *et al.*, 1988). Boron, a toxic element, is present in high concentrations (Mattigod, 1983). K is present in free ionic forms. Carbonates have also been extracted (Matusienwicz and Natush, 1980). A complex of organic components is also associated with fly ash particles. These are said to be adsorbed on to the surface of the particles.

Aromatic hydrocarbons (PAH), benzopyrene, benzofluranthenes, ideno (1,2,8 cd) pyrene, nitrophenanthrenes, and alkylated nitro fluorenones are present in fly ash (Hanson *et al.*, 1983; Harrison *et al.*, 1985; Low *et al.*, 1986; Harris *et al.*, 1987). Dimethyl sulphate, a non-aromatic mutagen/carcinogen has also been identified in coal fly ash (Eatough *et al.*, 1981). Ash pH varies from 4.5 to 12 depending on the S content of the parent coal. High S coal combustion resulted in acidic ashes whereas low S coal produced alkaline ashes during combustion (Singh and Singh, 1986).

Physical changes in soils due to fly ash amendments vary somewhat according to the original properties of soil and fly ash, but certain generalization can be made. Fly ash affects all physical characteristics of soil (Chang *et al.*, 1989). Soil-fly ash mixtures tend to have lower bulk densities, higher water holding capacities, lower hydraulic conductivities and lower moduli of rupture than soil alone. Although fly ash amendment increased water-holding capacity, it did not increase the amount of water available to plants (Chang *et al.*, 1977). Soils hydraulic conductivity improved at a low rate of fly ash treatment but when fly ash input increased (20%), it deteriorated, perhaps due to podzolanic reactions (Singh and Yunus, 2000). A significant cementing effect has been reported when fly ash was saturated with water or mixed with soil and water, especially if the soil was of sandy texture (Furr *et al.*, 1978 a; Weeldreyer and Fine, 1981). The cementing effects of fly ash could possibly impede root development by creating hard areas near the soil-fly ash interface resulting in reduced infiltration and root penetration in ash deposits and ash amended soils.

Since most of the fly ash have high concentration of alkaline earth elements it can be used to reduce the acidity of low pH soils. Elevated concentrations of trace elements found in fly ash can be used for overcoming soil nutrient deficiencies. During weathering many physico-chemical properties and elemental composition including both nutrients and inhibitory metals in fly ash, undergo changes (Jones and Lewis, 1960; Townsend and Hodgson, 1973) and this influences their effect on the environment and their possible-agronomic utilization as a liming agent or a micronutrient fertilizer (Mulford and Martens, 1970; Plank and Martens, 1974).

Fly ash brings about unfavorable changes in the soil's chemical equilibrium including increase in pH, salinity and levels of certain toxic elements. The effects of fly ash on chemical properties of soil are greatly influenced by the original pH of both the ash and the soil. Ash pH varies from 4.5 to 12 depending on S content of parent coal, however, generally it is alkaline with pH ranging from 8.2 to 12.5 (Furr *et al.*, 1977]. The greater alkalinity of fly ash is attributable to relatively high concentration of OH ions particularly those derived from lignite and sub-bituminous coal sources.

Soil salinity increases substantially when soils were amended with unweathered fly ash (Mulford and Martens, 1970; Adriano *et al.*, 1978: Elseewi *et al.*, 1978; Phung *et al.*, 1978; Page *et al.*, 1978). Fly ash addition in soil significantly increases the electrical conductivity (Singh *et al.*, 1994; Spencer and Drake, 1987; Khan *et al.*, 1996) *by* increasing the quantity of soluble major and micro nutrients along with the toxic heavy metal released by ash to soil (Eary *et al.*, 1990). Concentration of various elements in fly ash decreases with the increasing particle size, being greatest in clay sized fraction, less in the silt-sized fraction and least in the sand sized fraction (Davison et *al.*, 1974; Lee *et al.*, 1975; Natusch *et al.*, 1975; Campbell *et al.*, 1978; Adriano *et al.*, 1980). These elements are readily extractable in water (Linton *et al.*, 1976).

The most significant factors limiting vegetation establishment on ash deposits are excessive concentrations of soluble salts and boron. 30 mg/Kg of B is considered highly toxic to plants whereas fly ash contains above 200 mg/Kg of B (Bradshaw and Chadwick, 1980). High pH can also be a factor, limiting plant growth on ash. Fly ash pH depends on parent coal; high S coal produce acidic ash and low S coal produce alkaline ash (Adriano *et al.*, 1980; Bradshaw and Chadwick, 1980). Physical characteristics of ash deposits can also limit, successful vegetation establishment. Ash deposits develop concentrated or compacted layers that reduce aeration, water infilteration and root penetration.

Of all the possible uses of fly ash in a productive manner, the best use is as a fertilizer either through soil amendment or reclamation of fly ash dumps by cultivating plants (Anonymous, 1998). Efforts have been made to modify the properties of agricultural soil by addition of fly ash, thus improving the soil fertility and crop yield. Identification of plants capable of doing well in presence of fly ash has become necessary for getting sustained yields under the changed scenario.

Addition of fly ash to nutrient-deficient soil also have beneficial effects on crops. In some cases it may be due to a combination of pH adjustment and nutrients supplementation (Martens, 1971; Schnappinger *et al., 1975;* Elseewi *et al.*, 1978, 1980 b; Hill and Lamp, 1980). Fly ash improves crop growth by neutralizing the soil acidity (Moliner and Street, 1982). Besides, the potentially harmful chemical components of the ash are diluted by the soil in the soil-ash mixtures.

Fly ash disposal and utilization is an important area, of concern (Figure 1). Fly ash is also being utilized in construction industry mainly due to its podzolanic property (Gray and Lin, 1972; Manz, 1979; Schinley, 1979; Murtha *et al.*, 1983), it is being used extensively as an ingredient of cement, concrete mixes, and soil stabilizer in road construction and structural fills in highway construction. To encourage the use of fly ash in all construction activities, NTPC is setting up two fly ash bricks manufacturing plant, at Badarpur and Dadri near Delhi. Edil and Berthonex (1990) have used fly ash as earth liners for liquid waste containment Meanwhile, the DMRC (Delhi Metro Rail Corporation) has started using fly ash for construction of railway embankment and land fillings. The reduced bulk density of fly ash makes it a good material for lightweight blocks (Cope, 1962; Townsend and Hodgson, 1973; Phung *et al.*, 1978).

Fly ash is used to amend acidic mine spoils and mine wastes. Alkaline fly ash neutralizes, acidity, decreases bulk density, increases WHC and reduces compactness (Fail and Wochok, 1977; Jastrow *et al.*, 1981; Fail, 1987; Taylor and Schumann, 1988). Fly ash increases mine spoil pH, allowing the establishment of permanent vegetative cover (Capp, 1978). Fly ash being rich in Al and Fe, converts soluble P to insoluble compound and is used thereby to remove excess P from lakes and waste water (Fine and Jensen 1981; Weeldreyer and Fine, 1981). Fly ash being rich in metal oxides has been used as an ore for metal extraction (Gabler and Stoll, 1982; Gilliam *et al.*, 1982; Murtha *et al.*, 1983). In a major breakthrough fly ash has been converted into zeolite, a raw material for the detergent industry. It is the least expensive method of producing zeolite, an eco-friendly alternative to chemical additives in detergent (Anonymous, 1999).

Generation of electricity is treated synonymously with industrial growth and quality of life. Electric power is a very convenient source of energy, however its adverse environmental impacts vary depending on the methods of generation and the fuel used. In India there has been a rapid growth in power generation after independence (Table 2; Figure 2). The installed capacity has increased from 1,326 MW in 1947 to 76,800 MW by the end of 1994 (Singh and Yunus, 2000). Even with this impressive growth in the installed capacity our present per capita electricity consumption is nearly 280 kW hr^{-1} per annum in comparison with the developed countries, which is over 7.000 kW hr^{-1} per annum (Sohal, 1995).

Thermal power plants, are of course, a source of economic growth but these also bring a lot of environmental problems related to air, water and soil. The application of thermal power projects was motivated by the need to create a better social development and a sense of pride for the people but unfortunately at the same time they have become a threat to life. With such a disregard for the environment in the rush for power, the future cannot but be murkier, in the complete, sense of the word.

In the post-independence era, the power development has predominantly been through the state utilities. In the mid-seventies, it was realized that the efforts of the state utilities needed to be augmented by a central sector organization, which could contribute in rapid capacity addition by setting up large thermal power plants, in a time bound manner. To accomplish this gigantic task, NTPC was created in 1975. Today NTPC has an installed generating capacity of more than 14,6000 MW through its ten coal based thermal power projects located in Singruli (U.P.), Korba (M.P.), Ramagimdam (A.P.), Farakka (W.B.), Vindhyachal (M.P.), Rihand (U.P.), Kahalgaon (Bihar), Dadri (U.P.), Talcher (Orissa) and Unchahar (U.P.) and five gas-based power projects located at Anta (Rajasthan), Auriya (U.P.), Kamas (Gujarat), Dadri (U.P.) and Gandhar (Gujarat). In addition to it NTPC also manages 720 MW Badarpur Thermal Power Station (BTPS), which has the direct bearing on Delhi.

Delhi has three thermal power plants viz. Badarpur, Inderprastha and Rajghat. These three power plants on a whole emit an estimated 302 metric tones of pollutants every day (inclusive of particulate matter, SO_2, NO_x and hydrocarbons), however 63% of pollutants are released from BTPS alone (Source: Delhi Pollution Statistics, 1995). These pollutants apart from serious health hazards cause extensive damage to the vegetation in the vicinity of power plant.

BTPS is the main source of power supply to the city of Delhi (Plate 1). BTPS complex lies in the south of Delhi, near Delhi - Faridabad border. On the southern border of the town, flows the Agra canal, which is connected to the Yamuna River. The whole complex runs on the bituminous coal transported from various collieries of India. The chief chemical constituents of this variety of coal are 3-12% moisture, 15-40% ash, less than 10%, sulfur, 10-30% volatile matter, 20-40% carbon, 5-20% oxygen, 5% hydrogen and 5% nitrogen on an average. BTPS has 3 units of 100 MW each and 2 units of 210 MW each. The average daily coal consumption (in metric tones) is 1572, 1623, 1633, 3034 and 2577 from units I to V respectively and total daily average consumption is 10,439 tones (Figure 3). The total monthly and annual coal consumption from all the five units of BTPS is roughly estimated to be 3,20,000 and 37,60,000 metric tones respectively (Table 3).

The unburned coal possessing considerable amount of carbon, sulfur, hydrogen, nitrogen and oxygen, when subjected to high temperature for its combustion obviously produces CO_2, SO_2, NO_x and hydrocarbons. These gases spread into the atmosphere, through various stacks, as pollutants. Apart from the gaseous pollutants, pollution by particulate matter is also high enough due to high percentage of mineral matter and ash content of Indian coals. For every metric tonne of coal burned, 70- 80 kgs of fly ash is disposed off.

TABLE 1

Comparison of Properties of Coal, Fly Ash and Soil (Singh and Yunus, 2000)

Properties	*Coal*	*Fly Ash*	*Soil*
pH	—	Alkaline 8.2-12.8	6.7-7.6
Electrical conductivity (m mhos cm^{-1})	—	8.7	0.06-0.15
Cation exchange capacity ($mE1(100g)^{-1}$)	—	0.52-1.23	1.33-10.40
Organic carbon %	—	0.53-0.84	1.05-1.40
Total nitrogen %	—	0.0-0.02	0.02-0.01
Total phosphorus %	500	0.0-0.02	0.02-0.01
Exchangeable cations (mg g^{-1})			
Ca	5400	338-177100	0.7-50%
Mg	1200	116-60800	0.06-0.6%
K	1800	7360-22400	0.04-3.0%
Na	600	15-98	0.04-3%
Trace elements (mg g^{-1})			
Zn	39	20-152	10-300
Fe	16000	53-942	53-604
Ni	15	13-296	203-1000
Mn	100	29-145	11-14
Cu	19	75-132	2-200
Cd		52.37	40.50
Pb		94.56	36.26

TABLE 2

Decadal Growth of Thermal, Nuclear and Hydro Power Generation (MW)

	Thermal	*Nuclear*	*Hydro*	*Total*
1950	1005	0	559	1564
1960-61	2436	0	1917	4353
1970-71	7676	420	6383	14479
1980-81	17396	860	11791	30047
1990-91	44910	1465	18442	64817
1995-96	58675	1940	20512	81127

TABLE 3

Data on Coal Consumption (Metric Tones) in Five Units of Badarpur Thermal Power Station (BTPS), New Delhi

	Units of BTPS					
	I	*II*	*III*	*IV*	*V*	*Total Comsumption*
Annual	566068	584262	588021	1092207	927790	378348
Average monthly	47172	48689	49002	91018	77316	313196
Average Daily	1572	1623	1633	3034	2577	10439

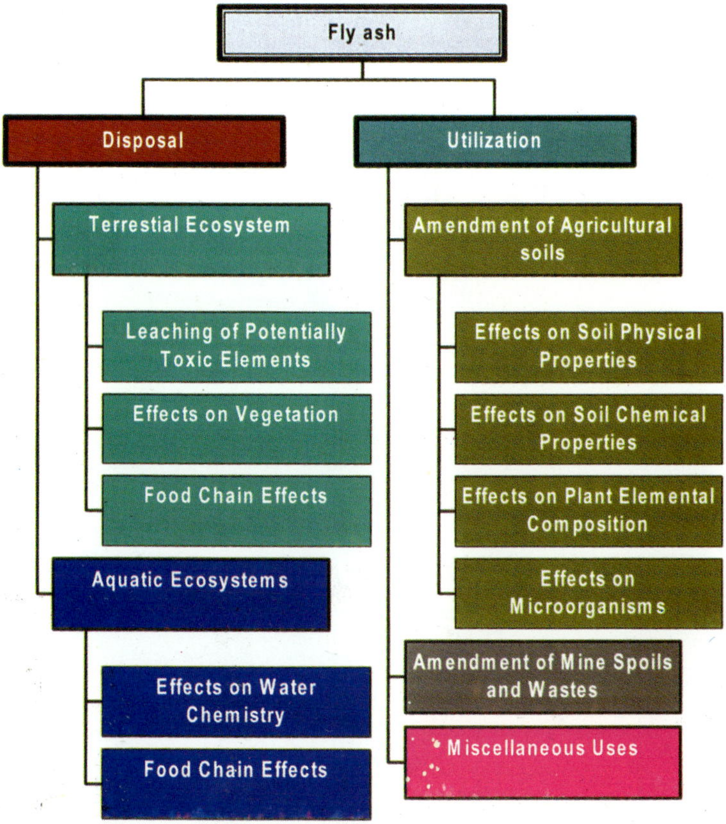

Figure 1: Schematic diagram of the potential environmental impacts of fly ash.

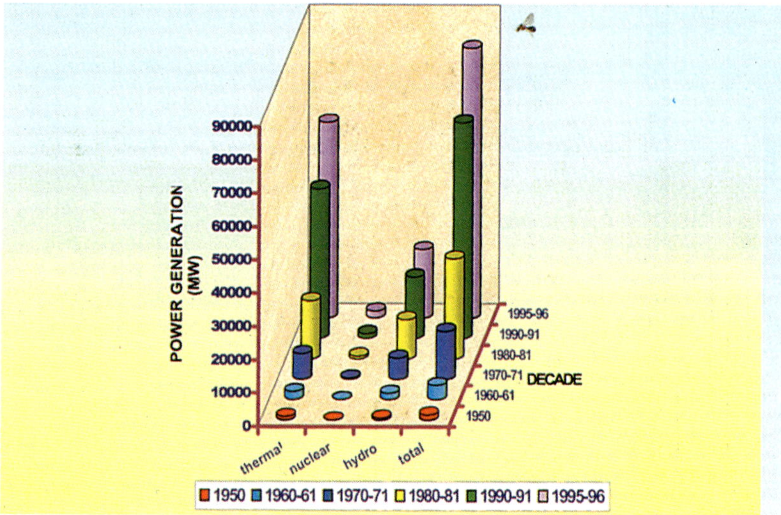

Figure 2: Comparision of decadal power generation (MW) in terms of thermal, nuclear and hydro power.

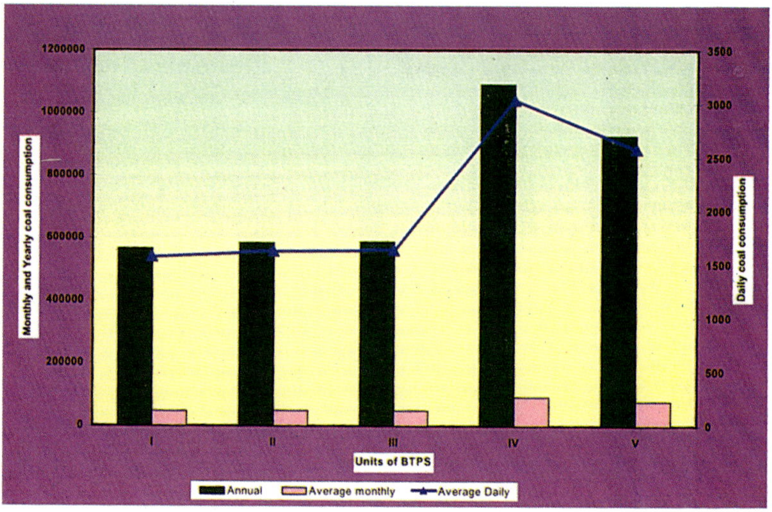

Figure 3: Daily, monthly and annual coal consumption (metric tonnes) in all the five units of BTPS.

2

Review of Literature: Fly Ash and Plant Growth

Ash Amendment of Soils and Response of Plants

Fly ash is not only a good growing medium, but also a better soil ameliorate (Hudson and Headley, 1990; Bulski *et al.*, 1995). Efforts have been made to modify the properties of agriculture soil by addition of fly ash, thus improving the soil fertility and crop yield. The effects on plants are primarily due to shifted chemical equilibrium induced by fly ash added to the soil. The varying edaphic factors and plant species further complicate interactions between plant and fly ash amended soil.

Fly ash is used to improve the texture of soil (Chang *et al.*, 1977, 1989) at different rates of application (Kuchanwar and Matte, 1997) by improving the water retention of the soils (Gangloff *et al.*, 1997). In black soils, fly ash addition decreased bulk density and increased porosity and aeration (Shinde *et al.*, 1995) resulting in increased groundnut yield at low rates of application (<7.5%), however at >10% level adverse effects on yield were reported. Similarly better seedling growth and total biomass of *Albizzia lebbeck* was observed by admixing fly ash with black soil and gypsum (Raizada *et al.* 1996) at low concentrations.

Brassica chinensis and *B. parachinensis* when grown in fly ash amended sandy and loam soils resulted in a high yield of both crops (Wong and Wong, 1990). Similarly fly ash application to sandy loam soils resulted in an increased crop yield of *Medicago sativa*, maize, muskmelons, potatoes and soybeans due to increased WHC and increased pH of soils amended with fly ash (Henning *et al.*, 1988).

Coal fly ash has been found to permanently improve the soil water relations of sandy, drought prone soils by changing infiltration rates (Gangloff *et al.*, 2000) and improving WHC of the soils (Salter and Williams, 1967; Chang *et al.*, 1989; Sharma et *al.*, 1990). Application of fly ash reduced soil crust strength from 2.38 kg/cm^2 to 0.98 kg/cm^2 in sunflowers and from 3.36 to 1.54 kg/cm^2 in groundnuts. This increased the yield of sunflowers from 0.60 to 0.76 t/ha and pod yield of groundnuts from 2.41 to 3.19 t/ha (Patil *et al.*, 1996). Fly ash is comprised primarily of silt and clay sized particles. It has the potential if applied at high enough rates to permanently change soil texture and increase moisture holding capacity (Ghodrati *et al.*, 1995, Sims *et al.*, 1995, Menzies and Aitken, 1996), however increased WHC did not generally improve plant yield. Chang *et al.* (1977) could not observe increase in plant yield as the water was held tightly by the soil so as to render it unavailable to plants. Salter *et al.* (1971) and Aitken and Bell (1985) hypothesized that low yield of crops even after increased WHC might be due to the presence of phytotoxic constituents in the ash. Whereas, Jacobs *et al.* (1991) did report significant increase in corn (*Zea mays* L.) yields with soil moisture increase by 25 to 70% during drought seasons.

Fly ash addition to the clayey soils significantly reduced WHC and bulk density contents of free lime and increased the content of fine sand which improved the soil texture thereby the growth and yield performance of sunflower (Kene *et al.*, 1991). Saxena *et al.* (1998) reported increased mobility of nutrients on mixing fly ash to clayey soil in addition to increased soil porosity and soil drainage.

Singh *et al.* (1994) observed that pH of the soil, amended with fly ash increased markedly. The electrical conductivity (EC) of the soil increased from 150-m mho cm^{-1} (for the control) to 416-m mho cm^{-1} at high application rates of fly ash. Considering the potential environmental impacts, fly ash can be used as a liming agent in acid soils by increasing pH and EC (Matsi and Keramidas, 1999). Similar reports of fly ash as an ameliorate for acidic soils thereby resulting in increased plant growth and crop yield have been extensively reported (Jiang *et al.*, 1999; Perkins and Vann, 1995; Raman *et al.*, 1996; Malewar *et al.*, 1999; Mc Leod and Ciravolo, 1997).

Fly ash application to acidic loamy soils resulted in increased dry weight and yield of *Elymus elongates* by increasing the soil pH from 6.3 to 8.25 (Wong 1995). Coreil (1997) reported that low rate of fly ash application increased significantly the yield of dryland cotton *(Gossypium hirsutum)*, however higher doses of fly ash adversely affected the growth of tomato, petunia and Boston fern due to very high pH of amended soil (Menzies and Aitken, 1996). Lower doses of fly ash favored the growth of *Albizzia procera* and *Acacia nilotica*, though higher rate of treatment resulted in retarded growth as against control unamended soils (Brahmachari *et al.*, 1994).

The major drawback of using fly ash as soil amendment is its chemical property. Increased alkalinity and salinity are generally the most immediate effects of ash application (Adriano *et al.*, 1982). Several studies have shown that addition of alkaline ash can increase the pH of acid soils resulting in adverse results (Plank *et al.*, 1975; Elseewi *et al.*, 1980; Moliner and Street, 1982; Elseewi and Page, 1984; Petruzzelli *et al.*, 1987). High pH and high soluble salt concentration are the limiting factors in the vegetation establishment on fly ash amended soils (Khan *et al.*, 1996).

Alkaline fly ash neutralizes acidity, decreases bulk density, increases WHC and reduces compactness (Fail and Wochok, 1977; Jastrow *et al.*, 1981; Taylor and Schumann, 1988). The cropping in fly ash amended and mine spoils increase the soil microbial biomass and activity leading to higher plant growth, biomass, yields and dry matter production (Bhumbla, 1992; Sajwan *et al.*, 1995; Lalk *et al.*, 1996; Reiberk and Kohrnak, 1984.). It is therefore used to amend acidic mine spoils allowing the establishment of permanent vegetative cover (Capp, 1978).

Fly ash contains little or no nitrogen (N) because the N originally present in the coal is volatilized during combustion (Bradshaw and Chadwick, 1980). Because of its deficiency in N and P, it is considered to be a low-grade fertilizer (Host and Pfenninger, 1978). Although fly ash generally has a high concentration of phosphorus (P) (400 - 8,000 mg P kg^{-1}), the form of P is not readily available to plants, presumably due to interactions with ash Al, Fe and Ca (in alkaline ash). Alkaline fly ash when added to fertilized soils increased the mean dry matter yields for 2% and 4% (w/w) treatments due to increased inputs of N, S and Fe from fly ash. However, at higher fly ash doses (8%), the yield was lowered due to lower P availability and possible toxicities of heavy metals (Sikka and Kansal, 1995).

Khan *et al.* (1996) observed an abrupt increase in the available P from their non-available form attributing it to the hydrolysis of Fe, Al and Mg compounds in fly ash at low concentration after prolonged watering in which release of inorganic acid is expected. These acids are subsequently used in the release of available P from their unavailable forms without affecting the pH of the system.

Fly ash does not contribute significant doses of plant macronutrients N, P and K to correct soil and plant deficiencies (Nass *et al.*, 1993; Padmakaran *et al.*, 1994; Somani *et al.*, 1994; Deshmukh *et al.*, 1999). Singh *et al.* (1991) reported that high yield of tomato and available N and K in soil amended with 500 t/h of fly ash that decreased significantly at 750 t/h of fly ash.

Similar observations were made for cotton yield and grain yield of wheat by Warambhe *et al.* (1992) and Gupta *et al* (1996) respectively. Application of 20% fly ash significantly increased the contents of N, P and K and their uptake in plants resulting in increased growth and yield of rice and wheat crops but >30% fly ash decreased the uptake of these nutrients thereby declining their growth and yield (Singh and Singh, 1986). The improved yields of potato, barley and rye have been reported at optimum rate of application due to improved soil structure and increased reserves of available Mg and K in soil (Katzur and Gora, 1986). Similar results of improved soil P, K and Mg contents and water retaining capacity of sandy soils, thereby increasing average yield, biomass and growth have been reported (Giedrojc *et al.*, 1980; Su, 1982; Podgiaskii and Mally, 1986; Wallace and Wallace, 1986; Thicke, 1989; Fox *et al.*, 1990).

Dry biomass yield of ryegrass (Matsii and Keramidas, 1999) and growth of spinach (Malewar *et al.*, 1999) significantly increased with fly ash application of acid soils. Similarly addition of fly ash in sandy soils as a replacement for P or K fertilizer increased the dry matter production of clover (Summers *et al.*, 1998). The plant height and yield of barley and sorghum crops increased in fly ash amended clayey soils as well, due to the increase in concentrations of available mineral nutrients in amended soils (Sale *et al.*, 1996; Sajwan *et al.*, 1996).

Efforts have been made to use fly ash in combination with sewage sludge to amend soil (Chaney *et al.*, 1976; Gutenmann *et al.*, 1981; Pietz *et al.*, 1989a, 1989b, 1989c; Sajwan *et al.*, 1995). However, the use of fly ash in agriculture has been based on its limiting potential and supply of nutrients such as Ca, B, S and Mo. Because of its deficiency in N and P, it is believed that the use of fly ash as an agricultural amendment can be enhanced by blending it with organic wastes such as sewage sludge which contains significant amount of N and P. Consequently, fly ash serves as a composting ingredients, along with sewage sludge (Sajjwan *et al.*, 1995). Addition of fly ash sludge mixture significantly improved the dry weight yields of *Agropyron elongatum*, but at higher, 35% and 50%, levels of application the dry weight yield was lower, though it was still higher than that of control (Su *et al.*, 1997). Certain microbes such as *Spirulina, Thiobacillus, Beggiatoa, Thiothrux, Thioploca etc.* serve as sources of biological N resulting in plant growth in fly ash amended soils with low N content (Rangaswami *et al.*, 1993; Banerjee and Deb, 1996). Fly ash generally contains sufficient concentration of elements but is deficient in Mn (Carlson and Adriano, 1991).

Because of the high concentration of many elements in fly ash, numerous studies have evaluated the usefulness of fly ash in nutrient-deficient soil. It can be used as a source of B (Plank and Martens, 1974; Wallace and Wallace, 1986), Ca (Martens and Beahm, 1976; Wallace *et al.*, 1980), Cu (Wallace *et al.*, 1980), K (Martens *et al.*, 1970), Mg (Hill and Lamp, 1980), Mo (Martens *et al.*, 1970; Cary *et al.*, 1983, Elseewi and Page, 1984), S (Elseewi *et al.*, 1978, Hill and Lamp, 1980) and Zn (Schnappinger *et al.*, 1975; Wallace *et al.*, 1980). Plants' uptake of K, Mn and Zn was increased with increasing percentages of fly ash resulting in increased growth and yield of *Zea mays* (Ghuman *et al.*, 1994). Observations complementing with increased uptake of metals resulting in increased yield at moderate rate of fly ash application have also been reported by Wong (1995).

Fly ash addition at 2% significantly increased the growth and yield of wheat (Petruzzelli, 1987). Similar results were reported in brome grass cv. Lincoln (Jastrow *et al.*, 1981) and *Brassica juncea, B. oleracea, Phaseolus vulgaris, Capsicum frutescens* and *Solanum melongena* (Menon *et al.*, 1993). Sims *et al.* (1995) reported that at high level of fly ash application, the available levels of metals also increased thereby resulting in declining effect on germination, seedling emergence and root shoot dry weight.

The seed germination of maize, sorghum, wheat and gram increased in soils treated with up to 10% fly ash and decreased with higher fly ash rates (Pawar and Dubey, 1988). Biomass production of *Agrostis tenius, Festuca arundinaceae* and *Lespedeza cerneate* ranged from 5 to 30 times high in fly ash treated plots as compared to untreated plots (Fail, 1987). Crops of *Amaranthus tricolor* and *Brassica chinensis* had higher yield in fly ash treated soil (Wong, 1990). Fly ash application resulted in higher enhanced vegetative growth and high tuber yield in the cutting of sweet potato cv BNAS-51 (Devaras *et al.*, 1982). The seedling development of tomato and *Capsicum annum* was found to be best in soil amended with fly ash at 25% level (Rengifo *et al.*, 1996).

High concentration of fly ash resulted in reduced germination viability and seedling growth, however at low to moderate rates, fly ash had a non significant increasing effect in maize and soybean (Mishra and Shukla, 1986). Ash amendments to the soil at low concentration increased grain yield of rice, but however caused physiological disturbance resulting in reduced grain yield at high concentrations (Machado *et al.*, 1984).

Good grain quality of autumn rye, *Medicago sativa,* barley and winter rape was achieved in various parts of ash deposits with the application of conventional agricultural techniques (Dzeletovic and Filipovic, 1995). The fly ash application increased the seed grain yield of kharif and rabi crops during their respective seasons of growth (Kuchanwar and Matte, 1997). Khandkar *et al.* (1996) reported high yield of rice, soybean and black gram *(Vigna mungo)* in fly ash amended soils. Enriching soils with fly ash increased the yield of groundnut and sunflower (Patil, 1996), onion *(Alium sativum)* (Chermisri *et al.*, 1995), wheat (Tripathy and Sahu, 1997), black currant (Kawecki *et al.*, 1993), soyabean and wheat

(Kumar *et al.*, 1999), pearl millet (Oswal *et al.*, 1999) and green gram *(Vigna radiata)* (Bhaisara *et al.*, 1999). Yield of barley silage was significantly increased at intermediate rates of fly ash application but significantly reduced at 400 t/ha (Hammermeister *et al.*, 1998).

Similar results of increased growth and yield of plants in soil and composts amended with fly ash have been reported (Davalos *et al.*, 1996; Mishra *et al.*, 1996; Raizada *et al.*, 1996; Scotti *et al.*, 1996; Masilamani and Dharmalingam, 1999; Prasad *et al.*, 2000).

Leaf number, number of branches, root weight, shoot weight and height of *Eycalyptus globolus, Syzigium cumini, Azadirachta indica and Annona squamosa* were affected significantly by the different fly ash and soil combination (Malewar *et al.*, 1998). Growth of lombardy *(Populus nigra)* and white poplars *(P. alba)*, common alder *(Alnus glutinosa)* and false acacia *(Robinia pseudoacacia)* were unsatisfactory on pure fly ash beds though most successful in gypsum and fly ash mixtures (Moffet and Shaw, 1993). Similarly growth measurements of *Eucalyptus tereticornis* and *Acacia auriculiforium* were best in fly ash amended soil and compost medium and poor in the only fly ash medium (Banerjee and Kashyap, 1998).

Sunflower *(Helianthus annus* L.) plants treated with fly ash exhibited improved growth (Pandey *et al.*, 1994). Relative growth rate (RGR) and net assimilation rate increased by over 20% at low fly ash application rate. Leaf area and leaf pigment content of the treated plants also increased. Similarly *Beta vulgaris* grown in fly ash-amended soil revealed that application of low amount (2%) of fly ash favored plant growth and improved yields (Singh *et al.*, 1994). Although metals such as Cd, Cu, Fe, Mu, Ni and Pb accumulated in plants, their levels remained well below the threshold limit and thus were suitable for human consumption. The increased sugar content at 2% fly ash amendment enhanced the possible use of fly ash in crop yield improvement. However, higher concentration of fly ash (4% and 8%) showed adverse effects.

Khan and Khan (1996) observed that tomato plant grown in the ash soil mixture showed luxuriant growth with bigger and greener leaves. Plant growth, yield, carotenoids and chlorophylls were enhanced in 40-80% fly ash amended soils. At 100% fly ash, yield was considerably reduced. The most economic level of fly ash incorporation was 40%, which improved the yield and market value of tomato fruit by 81 and 30% respectively.

Fly ash application at 20% reduced growth of tomato, petunia and Boston fern *(Nephrolepis exaltata)* while soil substrates containing 10% ash produced significantly increased growth of tomato (Menzies and Aitken, 1996). Fly ash applied to soil resulted in an increased seedling height, plant height, girth, leaf number, leaf area, spike length and dry weight of wheat at 5% rate of application (Tripathi and Sahu, 1997). Plant growth yield, carotenoid and chlorophyll contents of leaves were reported to be increased significantly as a result of fly ash application in chilly, egg plant and tomato (Khan and Ghadirpour, 1999). Foliar analysis of *Picea abies* grown in the vicinity of a thermal power plant, resulted in an increased ascorbic acid content, carotenoids and chlorophyll contents (Ribaric *et al.* 1999). Similar results of increased growth and leaf pigment content have also been discussed by Srivastava *et al.* (1995) due to fly ash application. *Phaseolus vulgaris* grown on fly ash amended soils resulted in a considerable increase in carotenoid content of leaves though chlorophyll a and b content were not greatly affected (Mercakchiiska *et al.*, 1995). Photosynthetic rate was also increased significantly with 20% fly ash enrichment in the soils.

The trace elements present in fly ash are potentially toxic due to their chemical limitations (Adriano *et al.*, 1980; El Mogazi *et al.*, 1988). These elements including As, B, Mo, Se and V are generally readily available to plant and tend to accumulate in plant tissues (Townsend and Hodgson, 1973; Hodgson and Buckley, 1975; Townsend and Gillham, 1975). High concentrations of elements in plants on fly ash amended soil may cause toxicity in animals grazing on them (Tolle *et al.*, 1983). Scanlon and Duggan (1979) planted eight species of trees and shrubs in dewatered fly ash landfills. Survival varied among species ranging from 12 to 84%. Analysis of foliar plant tissue revealed elevated levels of B, Ni and Se and in some cases As, Cd, Cu and Zn. In south California, the elevated levels of B, Cr, Co, Cu and V were reported in the trees grown on coal ash basin (Horton and Mc Minn, 1977; Mc Minn *et al.*, 1982). Carlson and Adriano (1991) found elevated trace element concentrations in sweetgum and sycamore growing on two abandoned coal fly ash basins. However B concentration remained within the tolerant range.

The soil when augmented with fly ash increases the amount of metal available to plants than those naturally occuring in soil (Straughan *et al.*, 1978; Cannon and Swanson, 1980; Elseewi and Page, 1984; Wadge and Hutton, 1987; Kuziemska *et al.*, 1989; Romer *et al.*, 1992; Gupta, 1993; Nass *et al.*, 1993; Warambhe *et al.*, 1993; Polyak *et al.*, 1994; Gupta and Gupta, 1998; Bernasconi *et al.*, 1999; Scotti *et al.*, 1999). However, over application of fly ash results in phytotoxicity, B being the most problematic (Adriano *et al.*, 1978; Elseewi *et al.*, 1981; Zwick *et al.*, 1984; Aitken and Bell, 1985). Plant tissue B contents were found to be increased according to rate of ash amendment and at high ash application rate, B toxicity symptoms in leaves were observed (Wong *et al.*, 1996; Kukier and Sumner, 1996; Hammermeister *et al.*, 1998). Cucumber plants grown on abandoned coal fly ash basin exhibited smaller leaves with leaf margin curls and necrosis resulting from toxicity of B, Mn and Zn (Dosskey and Adriano, 1993). Wong *et al.* (1998) observed injury symptoms due to B toxicity in leaf tips of *Elymus elongatus* at 35% and 50% ash amendment rates. Similar visible symptoms were evident in soil treatments with fly ash in Barley crops due to high concentration of B (Sale *et al.*, 1996). Increased rate of application of fly ash in soils resulted in much higher B content in roots and shoot causing marked reduction in plant growth (Kukier *et al.*, 1994). B as a dominant soil pollutant shows a negative correlation with yield. The plant growth and grain yield showed a high reduction to high B content in soil (Prausse, 1991) Ash amendments to soil resulted in elevated levels of B resulting in delayed seed emergence and reduced growth in *Zea mays* (Ghodrati *et al.*, 1995). However coal fly ash along with sewage sludge is a main source of B fertilizer and used effectively for high crop production (Gupta and Gupta, 1998; Hermann, 1996). Hence levels of fly ash must be determined for each growing medium to prevent B toxicity (Moree *et al.*, 1978).

Plants growing in fly ash areas accumulated high level of

Selenium (Se) resulting in Se toxicity. However, fly ash can be used an amendment to increase Se uptake by crops serving as an essential micronutrient for many deficient soils and crops (Shane *et al.*, 1988). Onions grown on coal fly ash contained a significantly high concentration of Se resulting in agricultural problems (Arthur *et al.*, 1992; Gutenmann and Lisk, 1996). Studies by Stoewsand *et al* (1978); Cary *et al.* (1983); Tolle *et al.* (1983) and Higgins and Fey (1993) indicate that plants grown on fly ash soils can contain 2 to over 200 tonnes as much Se as in plants on control soil leading to food chain risks.

The use of fly ash amendment in soils is considered a new method of nematode control resulting in better plant growth and improved yields (Haq, 1995) due to the presence of certain heavy metals in fly ash. Fly ash treatments significantly reduced the incidence of tikka leaf spot disease caused by *Meloidogyne berkeleyi* on the groundnut cv VRI-2 thereby leading to increased pod yield (Chandrasekar and Narayanswami, 1993). The root knot disease of tomato caused by the invasion of *Meloidogyne incognita* was adversely affected by fly ash treatment (Khan *et al.*, 1997). The disease intensity and reproduction of nematodes was decreased significantly thereby enhancing plants growth, leaf pigment concentration, fruit production, weight of fruit per plant and mean fruit weight (Ahmad and Alam, 1997).

Deposition of fly ash contaminate the soil, upsets the equilibrium of the ecosystem and causes genetic damage to flora in and around power plant. Fly ash contains heavy metals, which cause genetic damage to plants. Root of onion bulbs grown in fly ash polluted soil showed mitotic and chromosomal anomalies viz. clumping, stickiness, laggards and chromatin bridge formations. Fragments and c- mitosis appeared but in lower number due to heavy metal inhibition of spindle fibers (Sahi and Singh, 1996). The mitotic index of radish *(Raphanus sativus)* grown on soil amended with 10 to 50% fly ash showed an increase while at higher concentration (50-100%) of fly ash, clastogenicity and clumping of chromosomes was observed during mitotic divisions (Mishra *et al.*, 1999). Fly ash amended soil was used as growing medium for diploid and tetraploid maize crop but no triploid plants was observed at any fly ash concentration (Taets and Rayburn, 1996). Leaching of coal fly ash before admixing with soil reduces its genotoxic effect. Fly ash that had been weathered by leaching for one week did not cause changes in the mean DNA amount, and cell cycle (Mc Murphy *et al.*, 1996)

Fly Ash Deposition on Plants

Fly ash, under certain conditions of humidity stick to the leaves or fruits and promote chemical as well as physical injuries. Small necrotic dark brown spot appeared on leaves of many vegetables (green beans, turnip, cabbages, tomato etc.), vines and lemon trees near a fossil fuel burning power plant. Vaccarino *et al.* (1983) attributed it to the association of vanadium with fly ash. The vandate ions act as an inhibitor to the sodium pump (Macara, 1980) and negatively effect many enzymes (Catalan *et al.*, 1980). Fly ash and SO_2 synergistically affected the cucumber foliage by developing a specific mosaic chlorosis symptom on leaves (Singh and Mishra, 1988; Tung *et al.*, 1995). Leaf injury in *Azadirachta indica, Ficus glomerata* and *Nerium oleander* near powerhouse included tip necrosis and interveinal chlorosis (Śatyanarayanan and Pushpalatha, 1991). Fly ash particles may affect the mechanism that regulates the opening and closing of stomata, prevent their closure by blocking the stomatal aperture (Fluckiger *et al.*, 1979; Krajickova and Mejstrick, 1984) and thereby allow increased transpiration. The x-ray analysis and scanned electron microscopy (SEM) of *Picea abies* needles and *Fagus sylvestris* leaves confirmed the presence of trace elements (Si, Al, Mn, Fe etc.) in 98% of stomata in the vicinity of coal based thermal power plants producing fly ash (Mankovska *et al.*,1992).

Thick layers of fly ash also interfere with incidence of light and thus retard the photosynthesis (Mishra and Shukla, 1986). The length and density of trichomes increased whilst stomatal frequency and size decreased in *Commelina bengalensis* from the polluted environment (Mishra, 1982). The leaf area was decreased thereby retarding the rate of photosynthesis. Similarly reduced leaf blade size, stomatal frequency and guard cell size were observed in the studies with cocksfoot, tall oatgrass and Poa *annua* at sites close to fly ash source (Tietzeova and Mankusova, 1981). Substantially elevated concentrations of B were observed in crops dusted with fly ash. Elemental uptake occurs through intact leaf cuticles and stomata (Murray, 1984). Since fly ash contains more than 10% water soluble components, availability of B, S (Elseewi *et al.*, 1978); Ca, Mg (Plank and Martins, 1974; Adriano *et al.*, 1978) and Zn may be enhanced (Schnappinger *et al.*, 1975). The increased dry matter production is thus attributed to the increased availability of B through the leaf surface as observed in *Beta vulgaris, Cynadon dactylon* and *Medicago sativa* (Elseewi *et al.*, 1978); *Butea monosperma, Madhuca indica* and *Tectona grandis* (Dubey *et al.*, 1982).

Excessive uptake and accumulation of B, however hampers plant growth as noted in *Chloris gayana* (Mulchi and Armbruster, 1981) and *Phaseolus vulgaris* (Aitken and Bell, 1985). Fly ash deposition of 2 to 4 gm^{-2} day led to initial increase in transpiration rate, followed by a significant decrease when dusting rate increased to 8 gm^{-2} day^{-1} in maize and soybean cultivars (Mishra and Shukla, 1986). Increase in transpiration rate could be due to increased temperature of dusted leaves whereas the decreased rate in heavily dusted plants (8gm^{-2} day^{-1}) could be due to physical barrier created by the thick layer of fly ash.

The chlorophyll content of fly ash dusted leaves increased significantly (Dubey *et al.*, 1982; Pawar *et al.*, 1982; Mishra and Shukla,. 1986). Foliar analysis of *Picea abies* in the vicinity of thermal power plant resulted in an increased needle ascorbic acid content and photosynthetic pigments (Ribaric *et al.*, 1999). This could be due to shading effect caused by fly ash deposition (Mishra, 1982). However, this was not the only factor, because fly ash contains water-soluble salts (Rohrman, 1971) and leaf surface are known to absorb nutrients. This was thus reported to be cumulative effect (Iqbal, 2000). But in *Datura innoxia*, fly ash deposition on leaves resulted in low concentrations of photosynthetic pigments, though the levels of sugars, total phenols, free proline and the activity of oxidative enzymes was increased significantly (Satyanarayan *et al.*, 1988).

Reduction in chlorophyll content of high dusting rate was attributed to the alkalinity caused by excessive soluble salt on the leaf surface (Elseewi *et al.*, 1980) and increased foliar temperature which retards chlorophyll synthesis (Mark, 1963). Chlorophyll may be destroyed in the cells under a thick fly ash cover, as has been reported with reference to cement dust (Borka, 1980). It is also suggested that like ozone (Fletcher *et al.*, 1972; Adedipe *et al.*, 1973) or sulfur dioxide (Beckerson and Hofstra, 1979), coal particle pollution also causes induction of chlorophyllase or breakdown of chlorophyll to form pheophytin (Mudd, 1975). Similarly, photosynthesis was reported to be faster at low to moderate fly ash application but declined at increased dusting rate in maize and soyabean crops (Mishra and Shukla, 1986), possibly because increased foliar temperature accelerates photosynthetic activity at low rate of fly ash application, whereas a thick fly ash layer interferes with light.

Disposing off the huge amounts of fly ash is a major concern. Its use in manufacture of cement, concrete and bricks needs to be encouraged and requires research and developmental efforts with cost consideration. Changes in the soil characteristics and the cropping system due to addition of fly ash needs to be studied and investigated. It is desirable to establish vegetation on the landfill for stabilization against wind and water erosion thus providing shelter and habitat for wildlife and forming an aesthetically pleasing landscape (Hodgson and Buckley, 1975; Terman, 1978; Scanlon and Duggan, 1979; Adirano *et al.*, 1980; El Mogazi *et al.*, 1988; Singh *et al.*, 1992; Mishra *et al.*, 1995; Srivastava *et al.*, 1995; Petrikova *et al.*, 1996). Identification of vegetation capable of doing well in presence of fly ash particulate viz. air, water and soil has become necessary for getting sustained yields under the changed scenario.

Considering the potential environmental impacts of fly ash, the present study aims at investigating the influence of fly ash generated from Badarpur Thermal Power Station (BTPS) on two economically and medicinally important trees viz. *Cordia dichotoma* Forst f. and *Morus alba* Linn. Different concentrations (5%, 10%, 20%, 25% and 30%) of fly ash amendments in soils are used to ascertain the impact on both the species against the control-unamended soil.

3

Materials Used and Methods Involved

Cordia dichotoma Forst f. and *Morus alba* Linn, were the two trees selected for a comparative study of their performance in normal soil and soil amended with flyash at different concentrations.

PLANT DESCRIPTION

CORDIA DICHOTOMA FORST. F.

C. dichotoma belongs to family Boraginaceae and commonly known as Lasora. It is a small or medium sized tree with a short, usually crooked trunk 3-4ft. in girth. Fruit 0.5-1.0 inch long, yellowish brown, pink or nearly black when ripe, with a viscid, sweetish, almost transparent pulp surrounding a central stony part: branches glabrous: leaves stalked, alternate, entire or slightly dentate, rough when full grown, variable in shape from elliptic lanceolate to broad ovate, blade 3-6 inches long: flowers small, unisexual and bisexual on the same plant, whitish and in large open panicles.

Medicinal and Economic Importance

The fruit of *C. dichotoma,* popularly called sebestan, is edible, sweetish and mucilaginous. It is astringent, anthelmintic, diuretic, demulcent and expectorant. It is used in diseases of the chest and urinary tract infection. The kernels are used in external application against ringworm. A decoction of the bark is used in dyspepsia and fevers. The leaves are used for covering Burmese cheroots (Trotta, 1940), application on ulcers and in headaches.

The wood is clear yellow when freshly cut, durable in contact with water and gives a good surface finish. It is therefore used for boat construction, gum-stacks and agricultural implements.

The bark contains 2% tannin and is used in the tannin industry (Badhwar, 1944).It yields a fiber used for caulking boats, ropes and paper pulp (Anonymous, 1992).

MORUS ALBA LINN.

M. alba belongs to family Moraceae and commonly known as shehtoot and in english as mulberry. It is a monoecious, occasionally dioecious shrub or moderate sized tree with a fairly cylindrical straight bole, upto 3.0m. high and 1.8m. in girth. Bark dark grayish brown, rough with vertical fissures: leaves Very variable, ovate or broadly ovate, serrate or crenate serrate, often deeply lobed : flowers inconspicuous, greenish : male spikes (catkins) lax flowered, broadly cylindrical or ovoid, female spikes ovoid, pedunculate : fruit a syncarp consisting of many drupes enclosed in a fleshy perianth, ovoid or sub globose, upto 5cm. long, white to pinkish white, purple or dark purple to black.

M. alba is indigenous to China. It is extensively cultivated throughout the plains of India and in the hilly areas of Himalayas upto an elevation of 3,300mts. It is also grown as a road side and avenue tree.

Medicinal and Economic Importance

The fruits of *M. alba* are eaten fresh or made into juice and tarts; they may be squashed and fermented to yield spirituous liquors. The fruits contain a flavonoid, eriodictyol (Belavady and Balasubramanian, 1959; Ganju and Puri, 1959).

Mulberry leaves are considered diaphoretic and emollient. A decoction of leaves is used as a gargle in inflammations of the throat. The fruit is cooling and laxative; it is used for sore throat, dyspepsia and melancholia. The root is reported to possess anthelmintic and astringent properties. The bark is used as a purgative and vermifuge. Aqueous and alkali extracts of leaves and stems are active against Gram positive bacteria and yeast (Anonymous, 1962).

The wood of *M. alba* is much valued by the sports/goods industry on account of its elasticity and flexibility. Mulberry wood is used chiefly for hockey sticks, tennis and badminton rackets, cricket bat and stumps and other sports goods. The wood contains tannin; considered suitable for tanning and coloring purposes. The stem bark of *M. alba* is fibrous and has been employed for paper making. The twigs are used as binding material for making basket. Mulberry is grown extensively as leaves are used for rearing silkworms *(Bombyx mori* Linn.).

THE EXPERIMENTAL SETUP

Fly ash from Badarpur thermal power station (Plate l)was brought to the experimental site at the University campus. The fly ash and barren field soil were mixed. Experiments were conducted under natural conditions in fourteen-inch earthen pots having 10 kgs of varying soil and fly ash mixtures per pot. The fly ash doses were corresponding to *5%, 10%, 20%, 25%* and *30%* in the soil respectively. In the control pot no fly ash was added.

The pots were arranged in north- south direction in a randomized design for a possible uniform light condition. The saplings were suitably irrigated to provide a possible uniform soil moisture conditions. The sets of experiments were repeated for three complete years (March 1997- February 1998, March 1998- February 1999, March 1999- February 2000). Each year few weeks old saplings were transplanted to pots containing different soil mixtures in the month of February. The annual sampling was analyzed for three developmental stages of four successive months: I stage of leaf emergence and development (March-June), II stage of leaf maturity and optimum functioning (July-October), III stage of leaf senescence (November-February).

PARAMETERS STUDIED

Observations have been made on the following parameters in treated as well as control plants using established standardised protocols.

A. Morphological Parameters

(i) Foliar symptoms

(ii) Single leaf area

(iii) Leaf length

(iv) Leaf width

B. Anatomical Parameters

(i) Length and width of stomata

(ii) Length and width of stoma

(iii) Stomatal index

C. Physiological Parameters

(i) Net photosynthesis

(ii) Stomatal conductance

(iii) Intercellular carbon dioxide

D. Biochemical and Chemical Parameters

(i) Estimation of chlorophyll a and chlorophyll b

(ii) Estimation of total chlorophyll

(iii) Estimation of carotenoids

(iv) Estimation of nitrate

(v) Estimation of nitrate reductase activity

(vi) Estimation of total soluble protein

(vii) Estimation of soluble sugar

(viii) Estimation of sulfur

(ix) Estimation of nitrogen

(x) Estimation of pH

The data collected for the different parameters from the selected plants was statistically analysed with a view to find out the extent of variation and the degree of authenticity of the results.

4

Observations Recorded

The following results of experiments carried out on *Cordia dichotoma* and *Morus alba* in order to assess the impact of fly ash application on plant foliage are based on mean of the data recorded for annual sampling for three complete years (March 1997-February 1998, March 1998-February 1999, March 1999-February 2000). For the sake of convenience, different parameters are being dealt separately on a comparative basis and the annual cycle of sampling has been divided into three stages of four months each of plant growth: I stage of leaf emergence and development (March-June), II stage of leaf maturity and optimum functioning (July-October), III stage of leaf senescence (November-February).

Foliar Symptomology

Foliar symptoms and growth parameters were estimated at different plant developmental stages in control as well as treated plants. The growth parameters included leaf length, leaf width and single leaf area. The extent of the impact was estimated for different doses of fly ash in soil.

The leaves of *C. dichotoma* and *M. alba* grown on fly ash amended soil exhibited symptoms of injury at higher concentrations, though at lower concentrations the leaves had luxuriant growth. In *C. dichotoma* leaves the symptoms started showing up at 20% fly ash dose and further increased with the increase in fly ash concentrations (25% and 30%). The leaves showed development of slight interveinal chlorotic patches with the 20% treatment, which increased with 25% fly ash dose. The chlorotic patches increased in frequency and symptoms of leaf burn were evident at 25% level of fly ash treatment. With the increase in leaves chlorosis, slight necrotic patches were visible when 30% fly ash was added to the soil (Plates 2 and 3). The symptoms of foliar injury were more evident in *M. alba* grown on fly ash amended soil. Though there were no visible symptoms at 5% and 10% treatments of fly ash, but as the concentrations increased further (20%, 25% and 30%) the symptoms of injury started appearing on leaves. At 20% fly ash level, there was initiation of marginal leaf chlorosis with slight tip burn. Burning was also evident in leaf margins, which extended towards the middle vein. With the 25% dose of fly ash the patches of interveinal chlorosis started showing up and the tip burn extended to peripheral necrosis. The necrosis started intruding the entire lamina with 30% fly ash concentration in soil and the chlorotic patches turned necrotic, however interveinal chlorosis was still evident (Plates 4 and 5).

Foliar Morphology

Tables 4 to 7 present data on changes in single leaf area of both the plants in relation to fly ash addition at different stages of plant growth. The single leaf area was maximum at stage II for both the species. In *C. dichotoma* the leaf area increased at 5% fly ash dose and declined further with increasing dose of fly ash (10%, 20%, 25%, and 30%).

At 5% treatment of fly ash to the soil, single leaf area of *C. dichotoma* increased significantly with the maximum variation being (24.30%) in II stage of plant development. However a considerable decrease in single leaf area was recorded with increasing concentrations of fly ash, the maximum decrease (32.74%) being at 30% in the III stage (Tables 4 and 5; Figure 4). The data collected on *M. alba* showed highly significant increment in single leaf area at 5% and 10% dose of fly ash in soil. Though with further increasing concentrations a heavy loss in terms of single leaf area was recorded. Maximum variation in terms of increase (30.42%) and decrease *(27.68%)* was observed with 10% and 30% dose of fly ash, at III and II stages respectively (Tables 6 and 7; Figure 5).

Tables 8 to 11 present data on leaf length in relation to fly ash treatment in *C. dichotoma* and *M. alba.* The leaves attained a maximum length at stage II in both the species and for all the treatments of fly ash.

In *C. dichotoma* plants, leaf length decreased with increasing concentration of fly ash in the soil except at 5% level where the leaf length increased significantly than the control plants. The maximum variation in terms of decline (34.56%) was achieved with 30% fly ash level in the II stage whereas the major profit (48.47%) for leaf length was recorded with the 5% fly ash dose in the I stage (Tables 8 and 9; Figure 6). Almost similar trend was observed in *M. alba* fly ash treated plants, the only difference being at 10% level where the leaf length increased markedly. Here, the leaf length initially increased with the increase in fly ash level up to 10% but further decreased with the increasing concentrations of fly ash up to 30%. The variation in leaf length with 10% treatment was maximum at II stage (47.28%} and with 30% treatment, maximum loss of 26.37% was observed at II stage (Tables 10 and 11; Figure 7).

Changes in the leaf width of *C. dichotoma* and *M. alba* in relation to fly ash doses in soil are presented in tables 12-15. Leaf width was recorded maximum at stage II of plant growth for both the species.

In fly ash treated *C. dichotoma* plants, a significant increase in leaf width was recorded at 5% treatment with maximum variations (31.84%) being at stage I, followed by a decline with further increments in fly ash doses. The declining variation was maximum (29.77%) with 30% fly ash dose during stage II of plant development (Tables 12 and 13; Figure 8). In *M. alba* plants, the minor variations of 5% fly ash dose, became prominent with 10% dose of fly ash, showing the maximum decline (36.40%) at stage II. Leaf width further declined with increasing treatments of fly ash (30%), the degree of variation being the maximum (29.23%) at stage II (Tables 14 and 15; Figure 9).

Green Pigments in Leaves

Quantitative estimation of pigments i.e. chlorophylls and carotenoids, was made at different developmental stages of plant growth in control as well as treated *C. dichotoma* and *M. alba* plants. Tables 16 to 31 present data on changes in pigment concentrations with different doses of fly ash of both plant species.

Tables 16-19 present data on the total chlorophyll content as affected by fly ash treatments at different developmental stages of both the species. The total chlorophyll content declined in both the investigated trees with the increasing age. In control plants of *C. dichotoma* total chlorophyll increased in 5% fly ash treated plants, however with the increase in fly ash concentration in soil (10%, 20%, 25% and 30%) the total chlorophyll content showed a negative dose dependent relationship at each developmental stages. The total chlorophyll content reduced maximally (44.24%) with 30% fly ash treatment at stage 1 though a significant increment (44.42%) was observed with 5% fly ash dose at stage I of plant growth. (Tables 16 and 17; Figure 10). In fly ash treated *M. alba* control plants, the total chlorophyll content was more with 5% and 10% fly ash treatment, though there was a decrease in total chlorophyll content when 20%, 25% and 30% of fly ash was added to soil. The increase in total chlorophyll content was highly significant (73.11%) under the influence of 10% fly ash at stage II however the reduction was maximum (45.81%) with 30% fly ash in soil at stage III of plant growth (Tables 18 and 19; Figure 11).

Tables 20-23 present a comparative data on chlorophyll *a* content of *C. dichotoma* and *M. alba* with different fly ash treatments. Chlorophyll *a* content declined with the increasing age of both the tree species, the maximum being at stage I and minimum during stage III.

The chlorophyll a content in *C. dichotoma* plants showed a significant dose-dependent reduction at all stages of plant growth under the influence of fly ash treatments except for 5% dose, where it was a significant increment in chlorophyll a content. The maximum profit with 5% fly ash dose was recorded as 24.06% at stage III whereas the reduction in chlorophyll a content was highly significant with 30% fly ash treatment at stage II (46.15%). {Tables 20 and 21; Figure 12). In *M. alba* the maximum increment in chlorophyll *a* (63.91%) occurred at stage II with 10% fly ash treatment. The decrease was highly significant (39.48%) with 30% fly ash treatment at stage III of plant growth (Tables 22 and 23; Figure 13).

The comparison of chlorophyll *b* content in leaves of *C. dichotoma* and *M. alba* for different fly ash treatments during different stages of growth are presented in Tables 24 to 27.

The chlorophyll *b* content was greater in *C. dichotoma* plants treated with 5% fly ash, the maximum increment being 29.58% at stage II, followed by reduction with further increasing concentrations. The maximum reduction (38.52%) occurred with 30% fly ash treatment at the III stage. (Tables 24 and 25; Figure 14). The chlorophyll *b* content of *M. alba* increased at 5% and 10% doses of fly ash in soil and declined with further increased doses of fly ash (20%, 25%, and 30%). The reduction in chlorophyll *b* level was highly significant with 30% dose of fly ash (41.13%) at stage III of plant development, however the maximum increment of 79.03% was observed at 10% fly ash treatment at stage III (Tables 26 and 27; Figure 15).

The carotenoid content of leaves decreased as the plant age increased with each fly ash treatments given to both the investigated species. The carotenoid contents of *C. dichotoma* were relatively high at 5% fly ash dose as against control plants but a decline was observed as the fly ash dose in. the soil increased from 10 to 30%. (Tables 28 and 29; Figure 16). The decline in treated plants was highly significant (34.37%) at 30% fly ash level at stage III of growth but the maximum benefit (27.17%) in terms of carotenoid content occurred at stage I with 5% fly ash treatment. In *M. alba* leaves, the carotenoid content increased markedly in consequence of 5% and 10% fly ash treatments at all stages of growth except at the higher concentrations (20%, 25% and 30%) of fly ash where the carotenoid content dropped markedly, (Tables 30 and 31; Figure 17). The reduction was maximum (42.36%) at stage II with 30% fly ash dose and maximum increment (55%) was reported at stage III with 10% of fly ash in soil.

Foilar Sugar Content

Tables 32 to 35 contain data on concentration of soluble sugar in leaves of *C. dichotoma* and *M. alba* at all stages of development. The amount of sugar decreased with plant age in control as well as treated plants in both the species.

In *C. dichotoma* the sugar content increased initially with the lowest dose (5%) of fly ash and as the concentration of fly ash in soil increased {10%, 20%, 25% and 30%) it declined linearly. Maximum variations in terms of rise (43.05%) in sugar content was recorded at 5% level and in terms of loss (37.35%) was recorded at 30% level in stage 1 and stage III respectively (Tables 32 and 33; Figure 18). In *M. alba* the amount of sugar concentration in leaves enhanced with increasing percentage of fly ash in soil (5% and 10%), the maximum being at 10% level. Thereafter, as the concentration of fly ash in soil increased (20%, 25% and 30%) there was a sharp decline in sugar content, the maximum being at 30%. The maximum percentage variation when sugar level increased (55.31%) was recorded with 10% fly ash concentration and the maximum percentage variation in terms of loss (39.88%) in sugar level occurred at 30% fly ash concentration in soil at both stage III of plant growth (Tables 34 and 35: Figure 19).

Activity of Nitrate Reductase

Tables 36-39 shows the pattern of changes in NR activity with different fly ash treatments at specified developmental stages of both the tree species. The NR activity was found to be reduced at each developmental stage in treated as well as control plants.

In *C. dichotoma* plants, the NR activity showed a decreasing trend at all stages of plant growth with the increase in fly ash concentrations (10%, 20%, 25%, 30%) except at 5% treatment. The highest significant rise (69.24%) was observed at stage II with 5% concentration and the maximum reduction in variation (48.22%) was also observed at stage II, with the highest (30%) fly ash concentration (Tables 36 and 37; Figure 20). In *M. alba*, the NR activity showed a declining trend with the increasing plant age. The NR activity increased with. 5 and 10% fly ash levels in soil but on addition of 20%, 25% and 30% fly ash doses, the NR activity in leaves declined effectively. The maximum reduction (43.85%) was observed with 30% fly ash at stage II whereas the maximum significant rise (69.11%) was recorded with 10% fly ash at stage III of plant growth (Tables 38 and 39; Figure 21).

Foilar Nitrate Content

The leaf nitrate content in both the tree species increased with growing plant age. Tables 40 to 43 present data on leaf nitrate content at different developmental stages of plant growth in control as well as treated plants.

In *C. dichotoma,* nitrate content exhibited a maximum variation in reduction (27.09%) at 5% fly ash level and the most significant rise (82.05%) was found at 30% dose of fly ash in stage I of plant growth (Tables 40 and 41; Figure 22). With high fly ash concentrations, degree of increment in nitrate content in *M. alba* plants increased with growing age of the plants, however at low concentrations of fly ash (5% and 10%) there was a decline in leaf nitrate level. The maximum variation. (30.45%) in terms of decline was at 10% fly ash concentration in stage III and maximum significant increase (84.98%) was with 30% a fly ash dose in stage I of plant growth and development (Tables 42 and 43; Figure 23).

Total Soluble Protein in Foliage Tissues

Tables 44 to 47 present data on total soluble protein content in the leaves of *C. dichotoma* and *M. alba* at different concentrations of fly ash in the soil as against control unamended soil. The total soluble protein was estimated in the leaves of both the species at different developmental stages.

The total soluble protein content showed a decline with the increasing age of the plant in both the species.

In *C. dichotoma* plants, the total soluble protein content showed an increase with the addition of fly ash (5%) in the soil taut further addition of fly ash resulted in a decreasing trend of total soluble protein. However the most significant increment (57.71%) in protein level was observed with 5% fly ash at stage III of plant growth and the maximum loss (41.08%) in terms of protein content occurred at 30% during stage III (Tables 44 and 45; Figure 24). In *M. a*lba, the trend was almost similar to that of *C. dichotoma,* the only difference being that maximum protein content was observed at 10% dose of fly ash. The total soluble protein content increased at 5% and 10 % treatments of fly ash and declined on further increasing doses of fly ash (20%, 25% and 30%). The most significant variations for maximum profit. (69.15%) and maximum loss (41.46%) were obtained with 10% and 30% fly ash levels in soil at stage III and stage I respectively. (Tables 46 and 47; Figure 25).

Hydrogen Ion Concentration of Leaves

The leaf pH of *C. dichotoma* and *M. alba* exhibited a similar pattern with increasing plant age and increasing fly ash concentration. Leaf pH slightly increased with each stage of plant growth and development in both the tree species. Similarly with the increase in fly ash level in soil, the leaf pH showed a non significant increment. (Tables 48-51; Figure 26 and 27).

Elements in Foliar Tissues

Tables 52-55 present data on the leaf nitrogen content as affected by fly ash treatments at different developmental stages of *C. dichotoma* and *M. alba.*

The leaf nitrogen content of *C. dichotoma* was highest in the leaves at stage II and lowest during stage I of plant development in each fly ash treatments. At various fly ash treatments (10%, 20% 25% and 30%), nitrogen content of leaves showed a decline than control leaves except for 5% fly ash dose where it increased and was maximum. The similar trend was observed at each developmental stage. The variations in increment was maximum (9.54%) at 5% fly ash level, during stage III and in terms of decline (36.13%) was maximum at 30% dose during stage III (Tables 52 and 53; Figure 28). The nitrogen content for *M. alba* during different stages of development had a similar pattern as of *C. dichotoma*, the maximum being at stage II and minimum during stage I (Tables 54 and 55; Figure 29). With 5% and 10% level, N content increased in leaves and then declined with the increasing concentration (20%, 25% and 30%), the maximum being at 30%. The maximum percentage variation in terms of rise (12.42%) and decline (34.80%) occurred with 10% and 30% doses of fly ash during stage III and stage II respectively.

Tables 56 to 59 present data on the sulphur content in the leaves of *C. dichotoma* and *M. alba* during their successive growth in control as well as treated conditions.

In the leaves of *C. dichotoma*, the S content showed an irregular trend with plant age at different fly ash concentrations. However, with the increasing concentration of fly ash in soil, leaf sulphur content exhibited a declining trend. At low concentrations of fly ash the decline in S content was non-significant, however maximum percentage variation (16.87%) was recorded with the highest (30%) dose of fly ash in stage II of plant growth. (Tables 56 and 57: Figure 30). In *M. alba* leaves, the maximum sulphur content was observed at stage II and minimum S content at stage III with all doses of fly ash in soil. At stage I the S content in treated plants was less than that of control plants, however in stage II and stage III, the S content in the leaves of treated plants increased as compared to control plants. (Tables 58 and 59; Figure 31). The maximum decline (3.98%) in percent variation occurred with 30% fly ash concentration at stage I and the most significant variation in terms of increase (7.46%) was reported with 30% fly ash concentration at stage III of plant growth. The variations were however insignificant.

Photosynthetic Response

Tables 60 to 71, present data on physiological parameters such as net photosynthesis; stomatal conductance and intercellular carbon dioxide levels at different developmental stages of growth in control as well as treated plants of both the investigated species.

The rate of leaf photosynthesis decreased with plant age in

both the species. It was relatively reduced under each fly ash treatments except at 5%. fly ash dose in *C. dichotoma* leaves at every corresponding stage of plant development. The increase in net photosynthesis rate was highly significant (53.85%) with 5% fly ash treatment at stage II. This was followed by a decline in the rate of photosynthesis with the increase in the concentration of fly ash in soil, the maximum reduction (49.74%) being at 30% fly ash treatment at stage III of plant growth (Tables 60 and 61: Figure 32). Fly ash treatment to *M. alba* caused an increase in the photosynthesis rate at 5% and 10% treatments to soil, however as the concentration of fly ash in soil increased (20%, 25% and 30%), the net photosynthesis rate reduced significantly (Tables 62 and 63; Figure 33). In 10% fly ash treated plants, a highly significant rise (30.14%) occurred at III stage, while the photosynthesis rate got reduced with maximum variation (45.27%) also occurring at stage III under 30%. fly ash concentration.

In *C. dichotoma* plants, the stomatal conductance got retarded with increasing plant age. Higher concentration of fly ash (10%, 20%, 25%, and 30% had an inhibitory effect though 5% dose of fly ash showed positive results in terms of stomatal conductance. The decline with 30% treatment was maximum (43.75%) at stage III and the increase was maximum (61.84%) at stage II of plant growth with 5% fly ash dose (Tables 64 and 65; Figure 34). Similarly in *M. alba* plants, stomatal conductance declined with increased plant age. The stomatal conductance increased maximally, with 10% fly ash treatment (100.84%) at stage II and decreased maximally (54.62%) with 30% fly ash at stage II of plant growth (Tables 66 and 67; Figure 35). Though an increase was observed with 5% and 10% fly ash in soil but at further high concentrations a declining dose dependent relationship was observed.

Intercellular CO_2 concentration in leaf tissue was maximum during stage III and was minimum during stage I in both the tree species. The intercellular CO_2 concentration of *C. dichotoma* leaves increased with the increasing concentration of fly ash except at 5% where it was minimum. The increase was maximum (48.37%)

at stage I with 30% dose of fly ash in soil while maximum decline (25.89%) was observed at stage II with 5% of fly ash (Tables 68 and 69; Figure 36). The effect of fly ash treatment on intercellular CO_2 conc. was almost similar in *M. alba* plants. It showed an increase with the increasing plant age. The maximum increase (22.55%) occurred at stage I with 30% of fly ash dose though maximum variation in reduction (22.57%) was recorded with 10% fly ash dose at stage I (Tables 70 and 71; Figure 37).

Foliar Anatomy in Relation to Epidermal Responses

The data collected on epidermal traits at different developmental stages of plant growth of *C. dichotoma* and *M. alba* under the influence of fly ash treatment are discussed here under. The leaves were hypostomatic in *C. dichotoma* however the leaves of *M. alba* bear stomata only on the abaxial epidermis. The changes in size of stomata and stoma as well as stomatal index decreased with increasing age of plants in both control as well as treated samples of both the species. However, with the increase in concentration of fly ash these parameters experienced an irregular trend.

Stomatal index on the abaxial epidermis of control plants of *C. dichotoma* was maximum at stage I and decreased with the increasing plant age at all fly ash treatments. Fly ash had an irregular dose dependent effect: on stomatal index at different concentrations. Initially, the stomatal index increased with the increase in 5%, 10% and 20% concentration of fly ash in soil, but the maximum significant increase (17.66%) was recorded with 5% level of fly ash at stage II. However, as the concentration of fly ash rose to 25% and 30% there was a dose dependent inhibitory effect at each stage of plant development. The reduction in stomatal index was maximum (19.39%) with 30% treatment during stage II (Tables 72 and 73; Figure 38). In *M. alba* highest stomatal index on abaxial epidermis was recorded at stage I and lowest at stage III of plant growth and development. However, an increase in stomatal index was recorded with 5% and 10% fly ash concentration in soil, which declined with further increasing concentration (20%, 25% and 30%) of fly ash in soil (Tables 74 and 75; Figure 39). The maximum significant variation in terms of increased stomatal index (15.60%) and decreased stomatal index (16.37%) occurred with 10% and 30% dose of fly ash during stage I and stage III respectively.

Stomatal index on adaxial epidermis of *C. dichotoma* decreased with the increase in plant age at each fly ash doses. However, with all the fly ash treatments stomatal index decreased except for 5% level which experienced a significant rise (28.24%) in stomatal index at stage III. The lowest stomatal index was recorded with 30% fly ash level, the maximum variation (51.45%) being at stage III. (Tables 76 and 77; Figure 40).

The length of stomata on the abaxial epidermis was maximum at stage I and minimum at stage III of plant growth for both the tree species. In fly ash treated *C. dichotoma* plants, the stomatal length increased with 5%, 10% and 20% doses of fly ash, the maximum increase being at 5% level. Thereafter It 25% and 30% fly ash levels, a decline in stomatal length was observed (Plates & and 7). The maximum increase in percent variation in terms of increase (30.31%) and decrease (33.30%) of stomatal length occurred at stage II and stage III with 5% and 30% treatment of fly ash in soil respectively. (Tables 78 and 79; Figure 41). In fly ash treated *M. alba* plants the length of stomata on abaxial epidermis reduced significantly except at a few lower concentrations. The stomatal length increased linearly with fly ash concentrations up to 10% level, later with the increasing concentrations (20%, 25% and 30%) a decreasing pattern of stomatal length was observed. The maximum increase (28.07%) in stomatal length was recorded for 10% fly ash at stage III while the maximum significant decrease (29.93%) of stomatal length occurred with 30% fly ash at stage I of plant growth (Tables 80 and 81; Figure 42).

In *C. dichotoma,* the stomatal length of adaxial epidermis declined almost consistently with the advancement in plant age, with the increasing concentrations of fly ash (10%, 20%, 25% and 30%) except at the lowest concentration (5%) where length of stomata was recorded to be maximum in size. The maximum variation in terms of increase (9.29%) and decrease (26.99%) in stomatal length was observed at 5% and 30% treatments of fly ash at stage I and stage II respectively (Tables 82 and *83;* Figure 4.3).

In fly ash treated *C. dichotoma* plants, the stomatal pore length on abaxial epidermis was maximum at stage I and decreased with further increase in age. The length of stoma increased at 5%, 10% and 20% levels but when the fly ash dose increased up to 25% and 30% a highly significant decline was evident at all stages of plant growth (Plates 8 and 9). The highest percent variation in terms of increase (37.47%) in length of stomatal aperture was recorded at 5% during stage I and the significant decline (39.86%) in pore length was observed during stage III *with* 30% fly ash dose. (Tables 84 and 85; Figure 44). The stomatal pore length on abaxial epidermis decreased with the increasing age of plant at all concentrations of fly ash in *M. alba.* The increment in length of stomatal aperture was evident only with 10% fly ash treatment and for the rest of the treatments, a decline in pore length was evident at all stages of growth. The percent variation was significant for maximum increase (24.40%) and maximum decrease (42.91%) in pore length at 10% and 30% treatments of fly ash during stage II and stage III respectively (Tables 86 and 87; Figure 45).

The variations on adaxial epidermis of *C. dichotoma* were less significant as compared to abaxial epidermis. Stomatal pore length was maximum at stage I and eventually declined to a lowest during stage III. With the increasing concentration of fly ash, a decreasing trend in length of stomatal aperture was recorded. The decline in stomatal pore length was maximum (37.86%) at 30% fly ash concentration during stage III (Tables 88 and 89; Figure

The width of stomata on abaxial epidermis of fly ash treated *C. dichotoma* increased at 5%, 10% and 20% levels but got reduced at 25% and 30% levels as compared to control plants for all stages of plant growth. The variation in increment was maximum (43.65%) at 10% level of fly ash in soil at stage III. The maximum significant percent variation in terms of reduction (28.17%) occurred with 25% fly ash dose at stage III of plant growth (Tables 90 and 91; Figure 47). In *M. alba* plants, stomatal width on abaxial epidermis increased consistently with the increase in fly ash concentration in comparison to control plants at all stages of plant growth (Plates 10 and 11). The maximum percent variation for the significant increase (53.09%) was reported during stage III with 10% treatment of fly ash (Tables 92 and 93; Figure 48).

In *C. dichotoma,* the width of stomatal aperture on adaxial epidermis declined with increasing plant age for all the fly ash treatments. Width size was more for all the concentrations of fly ash as against control plants grown on unamended soil. The maximum percent variation (73.11%) in terms of increase in stomatal width was recorded with 5% dose of fly ash during stage I of plant growth (Tables 94 and 95; Figure 49).

The width of stomatal aperture on abaxial epidermis decreased with increasing plant age in *C. dichotoma,* with the maximum size during stage Stomatal pore width in fly ash treated plants was more than that of control plants for 5%, 10% and 20% doses of fly ash however for 25% and 30% treatments the stomatal width decreased. The maximum reduction (75.59%) in width was observed with 25% treatment at stage III and maximum increment (47.89%.) in stomatal pore width occurred with 10% level of fly ash at stage II of plant growth (Tables 96 and 97; Figure 50). In *M. alba,* also the size of stomatal width on abaxial epidermis decreased with the increasing plant age at all fly ash doses. A significant increase in the width of stomatal aperture was observed with almost all the fly ash treatments at each stage of growth (Plates 12 and 13;. The increase in size was maximum (47.61%) with 10% treatment however decrease in size was maximum (43.62%) with 30% at stage III of growth and development (Tables 98 and 99; Figure 51).

On the adaxial epidermis of *C. dichotoma,* the decline in stomatal pore width was evident with all the fly ash treatments as the plant ages (Tables 100 and 101). The reduction in size was however significant only with higher concentration of fly ash, the maximum (52.88%) being at 30%, during stage II (Plates 14 and 15; Figure 52).

Soil Analysis

The soil pH was measured for all the soil-fly ash doses at the first month and the last month of sampling of both the tree species. The pH of fly ash treated soils increased with the increasing concentrations of fly ash, though the increase was non-significant (Table 102). The pH declined insignificantly at the end of sampling for both the species during all the fly ash treatments. The soil N increased insignificantly with the increase in fly ash concentration in soil. Nitrogen level in soil was reported to be reduced at the end of final sampling insignificantly (Table 103). Table 104 shows that there was no specific change in sulphur concentration of different fly ash amended soil however at the end of the final sampling the sulphur level in soil declined than the initial level at the start of sampling.

TABLE 4

Single Leaf Area (cm²) *of C. dichotoma* Growing on Different Concentrations of Fly Ash Amended Soil During an Annual Cycle of Three Consecutive Years

	Control	*5%*	*10%*	*20%*	*25%*	*30%*
March	28.43 ± 5.91	32.98 ± 3.96 (16.04)*	27.42 ± 6.67 (3.5)	26.59 ± 4.28 (6.47)	22.15 ± 3.53 (22.09)*	21.99 ± 2.07 (22.65)*
April	31.36 ± 4.67	36.71 ± 4.19 (17.06)*	32.58 ± 4.79 (3.89)	28.27 ± 3.62 (9.86)	24.32 ± 3.15 (22.49)*	22.75 ± 1.95 (27.46)**
May	33.28 ± 2.05	38.28 ± 3.07 (15.02)*	33.33 ± 1.42 (0.45)	31.19 ± 1.89 (6.28)	26.84 ± 2.01 (19.35)*	25.28 ± 4.65 (24.03)*
June	33.68 ± 2.85	42.97 ± 2.43 (20.21)*	34.25 ± 4.76 (4.07)	32.47 ± 2.24 (8.96)	26.84 ± 3.67 (25.19)*	25.28 ± 3.06 (26.74)**
July	36.71 ± 3.07	43.49 ± 4.17 (18.47)*	36.47 ± 3.12 (0.65)	32.65 ± 3.06 (11.06)	28.15 ± 4.57 (23.31)*	27.79 ± 3.65 (24.29)*
August	36.34 ± 5.43	45.29 ± 5.73 (24.62)*	36.98 ± 3.03 (1.765)	33.05 ± 4.42 (9.05)	28.95 ± 3.75 (20.36)*	27.72 ± 2.14 (23.72)*
September	37.19 ± 9.85	46.62 ± 6.18 (25.36)*	38.71 ± 6.42 (4.09)	33.89 ± 5.32 (8.87)	29.26 ± 8.98 (21.32)*	28.29 ± 3.17 (23.93)*
October	36.16 ± 6.33	46.58 ± 8.68 (28.82)**	37.12 ± 5.35 (2.65)	34.34 ± 5.27 (5.04)	29.02 ± 4.65 (19.75)*	28.86 ± 4.07 (20.19)*
November	34.03 ± 8.96	44.28 ± 8.79 (30.12)**	32.44 ± 4.34 (4.67)	30.54 ± 4.25 (10.26)	27.83 ± 3.15 (18.21)*	25.433 ± 6.89 (25.27)*
December	32.12 ± 4.39	41.52 ± 6.24 (29.27)**	31.23 ± 3.06 (2.77)	28.41 ± 4.98 (11.54)	26.28 ± 3.82 (18.17)*	22.14 ± 5.87 (31.07)**
January	30.43 ± 3.59	35.23 ± 3.27 (15.77)*	29.14 ± 2.19 (4.24)	25.33 ± 2.46 (16.76)*	23.06 ± 1.94 (24.21)*	18.15 ± 1.26 (40.35)**
February	31.42 ± 3.59	35.58 ± 3.17 (13.24)	30.79 ± 2.19 (2.06)	26.98 ± 2.46 (14.11)	25.51 ± 1.94 (18.89)*	20.37 ± 1.26 (35.17)*

* = sig at 1% ** = sig at 5% (for tables 4-104)

TABLE 5

Single Leaf Area (cm²) of *C. dichotoma* Growing on Different Concentrations of Fly Ash Amended Soil During Different Stages of Growth

	Control	*5%*	*10%*	*20%*	*25%*	*30%*
Stage 1	32.18 ± 3.37	37.71 ± 3.41 **(17.17)***	31.89 ± 4.41 **(0.91)**	29.63 ± 3.01 **(7.94)**	25.00 ± 3.09 **(22.32)***	24.04 ± 2.93 **(25.31)***
Stage 2	36.6 ± 6.17	45.49 ± 6.19 **(24.30)***	37.32 ± 4.48 **(1.96)**	33.28 ± 4.51 **(8.52)**	28.48 ± 5.48 **(21.18)***	28.16 ± 3.25 **(23.04)***
Stage 3	32.00 ± 5.69	39.15 ± 6.01 **(22.35)***	30.9 ± 3.55 **(3.44)**	27.81 ± 3.74 **(13.07)**	25.67 ± 2.81 **(19.78)***	21.52 ± 4.03 **(32.74)***

TABLE 6

Single Leaf Area (cm^2) of *M. alba* Growing on Different Concentrations of Fly Ash Amended Soil During an Annual Cycle of Three Consecutive Years.

	Control	*5%*	*10%*	*20%*	*25%*	*30%*
March	26.69 ± 5.74	33.37 ± 4.29 **(12.39)**	39.02 ± 5.36 **(31.43)****	27.25 ± 5.62 **(8.21)**	24.51 ± 2.86 **(17.44)***	24.12 ± 2.06 **(18.59)***
April	30.86 ± 51.2	33.62 ± 6.69 **(8.92)**	39.45 ± 6.83 **(27.83)****	28.94 ± 2.72 **(6.22)**	26.06 ± 368 **(15.56)***	25.23 ± 2.39 **(18.24)***
May	32.33 ± 7.69	36.45 ± 6.69 **(13.09)**	40.49 ± 5.15 **(24.38)***	28.62 ± 6.29 **(14.23)**	27.18 ± 5.84 **(15.91)***	25.24 ± 5.11 **(21.68)***
June	34.14 ± 9.99	39.24 ± 8.86 **(14.93)**	41.43 ± 7.64 **(21.35)***	29.07 ± 6.15 **(11.12)**	27.69 ± 7.99 **(15.94)***	25.64 ± 5.49 **(22.95)***
July	35.25 ± 5.87	39.48 ± 8.49 **(12.43)**	43.98 ± 9.21 **(24.7)***	31.33 ± 6.72 **(11.12)**	29.63 ± 6.51 **(15.94)***	27.16 ± 7.22 **(22.95)***
August	38.01 ± 4.23	41.65 ± 5.17 **(9.83)**	47.14 ± 5.29 **(21.96)***	34.55 ± 4.28 **(12.05)**	31.03 ± 3.48 **(25.54)****	27.29 ± 3.94 **(28.96)****
September	39.98 ± 9.59	43.63 ± 7.65 **(8.45)**	47.98 ± 6.01 **(20.07)***	35.09 ± 8.11 **(12.23)**	30.24 ± 3.92 **(24.36)***	27.63 ± 8.62 **(30.89)****
October	36.55 ± 5.27	40.68 ± 5.91 **(8.11)**	44.52 ± 6.14 **(23.26)***	32.27 ± 6.33 **(11.10)**	29.53 ± 4.59 **(18.21)***	26.97 ± 4.84 **(27.44)****
November	34.83 ± 6.11	37.43 ± 7.81 **(7.25)**	41.18 ± 5.81 **(17.72)***	31.29 ± 8.71 **(9.87)**	25.69 ± 969 **(25.51)****	24.23 ± 4.71 **(29.58)****
December	25.25 ± 2.43	31.14 ± 2.36 **(23.32)***	22.35 ± 4.40 **(36.36)****	25.71 ± 3.66 **(1.82)**	19.78 ± 4.78 **(13.74)**	18.64 ± 3.65 **(28.25)****
January	22.29 ± 47.69	28.15 ± 5.64 **(26.28)****	30.66 ± 6.11 **(46.52)****	21.75 ± 5.89 **(2.46)**	17.73 ± 6.18 **(11.99)**	16.65 ± 4.89 **(26.33)****
February	28.12 ± 5.11	29.11 ± 6.18 **(3.52)**	36.22 ± 8.15 **28.84)****	24.81 ± 5.72 **(11.77)**	22.28 ± 4.11 **(20.76)***	18.43 ± 3.97 **(30.72)****

TABLE 7

Single Leaf Area (cm^2) of *M. alba* Growing on different Concentration of Fly Ash Amended Soil During Different Stages of Growth

	Control	*50%*	*10%*	*20%*	*25%*	*30%*
Stage 1	31.73 ± 7.18	35.67 ± 6.7 **(12.42)**	39.99 ± 6.24 **(26.05)****	28.47 ± 5.195 **(10.27)**	26.36 ± 5.09 **(16.92)***	25.07 ± 3.76 **(20.99)***
Stage 2	37.69 ± 6.24	41.29 ± 6.78 **(9.53)***	46.15 ± 6.66 **(22.43)**	33.31 ± 6.36 **(11.64)**	30.11 ± 4.62 **(20.13)***	27.26 ± 6.15 **(27.68)***
Stage 3	27.37 ± 4.58	31.71 ± 5.49 **(13.76)**	36.35 ± 6.12 **(30.42)****	26.13 ± 5.99 **(6.22)**	22.87 ± 6.19 **(17.92)***	21.00 ± 4.34 **(24.65)***

TABLE 8

Leaf Length (cm) of *C. dichotoma* Growing on Different Concentrations of Fly Ash Amended Soil During an Annual Cycle of Three Consecutive Years

	Control	*5%*	*10%*	*20%*	*25%*	*30%*
March	6.09 ± 2.35	7.58 ± 1.39 **(24.46)***	6.71 ± 2.07 **(10.18)**	5.64 ± 1.23 **(7.39)**	5.08 ± 2.57 **(16.58)***	4.54 ± 1.69 **(25.45)***
April	6.68 ± 2.85	10.69 ± 2.12 **(55.83)****	7.24 ± 2.13 **(5.53)**	6.63 ± 1.58 **(3.35)**	5.45 ± 1.92 **(20.55)***	5.22 ± 1.71 **(23.97)***
May	7.12 ± 2.31	11.54 ± 2.95 **(62.08)****	6.77 ± 2.49 **(4.92)**	6.71 ± 1.15 **(5.75)**	6.07 ± 2.67 **(14.74)**	5.29 ± 1.42 **(25.71)***
June	8.03 ± 1.57	11.91 ± 2.32 **(48.32)****	7.93 ± 1.14 **(1.24)**	6.98 ± 2.46 **(13.07)**	6.21 ± 1.74 **(22.61)***	5.32 ± 1.39 **(33.74)***
July	8.15 ± 3.41	12.77 ± 2.92 **(56.68)****	8.71 ± 2.11 **(6.87)**	7.25 ± 1.91 **(11.04)**	6.27 ± 1.45 **(23.07)***	5.46 ± 1.24 **(33.56)****
August	9.71 ± 2.06	13.15 ± 3.96 **(35.42)****	9.54 ± 2.96 **(1.75)**	7.31 ± 2.12 **(24.72)***	7.25 ± 1.15 **(25.33)***	6.08 ± 2.51 **(37.38)****
September	9.86 ± 4.76	13.67 ± 3.14 **(38.64)****	9.75 ± 4.16 **(1.12)**	8.24 ± 1.98 **(16.43)***	7.17 ± 2.28 **(27.28)****	6.22 ± 2.87 **(36.91)****
October	9.37 ± 2.78	13.22 ± 2.13 **(41.08)****	9.68 ± 1.26 **(3.38)**	8.35 ± 2.84 **(10.89)**	6.87 ± 1.76 **(26.68)****	6.51 ± 1.86 **(30.52)****
November	8.28 ± 2.62	11.21 ± 2.75 **(35.38)****	7.99 ± 1.93 **(3.52)**	6.91 ± 1.83 **(16.55)***	6.23 ± 2.23 **(24.75)****	5.94 ± 1.21 **(28.26)****
December	7.23 ± 2.95	10.91 ± 2.04 **(50.89)****	6.03 ± 1.19 **(16.59)***	6.32 ± 1.51 **(12.58)**	5.12 ± 1.62 **(29.18)****	5.02 ± 1.59 **(30.56)****
January	6.12 ± 2.53	8.35 ± 1.82 **(36.43)****	5.52 ± 1.76 **(9.83)**	5.38 ± 2.98 **(12.09)**	4.81 ± 1.32 **(21.44)***	4.47 ± 0.73 **(26.96)****
February	6.41 ± 1.41	9.31 ± 2.54 **(45.24)****	5.87 ± 1.98 **(8.42)**	5.12 ± 2.25 **(20.12)***	4.97 ± 1.63 **(24.46)***	4.11 ± 1.07 **(39.58)****

TABLE 9

Leaf Length (cm) of *C. dichotoma* Growing on Different Concentrations of Fly Ash Amended Soil During Different Stages of Growth

	Congrol	*5%*	*10%*	*20%*	*25%*	*30%*
Stage 1	7.02 ± 2.27	10.43 ± 2.19 **(48.47)****	7.16 ± 1.95 **(1.95)**	6.49 ± 1.60 **(6.61)**	5.70 ± 1.97 **(18.82)***	5.09 ± 1.55 **(27.51)***
Stage 2	9.27 ± 3.25	13.20 ± 3.03 **(42.38)****	9.42 ± 2.62 **(1.59)**	7.78 ± 2.21 **(16.01)***	6.89 ± 1.66 **(25.69)****	6.06 ± 2.12 **(34.56)****
Stage 3	7.01 ± 2.37	9.94 ± 2.28 **(41.86)****	6.35 ± 1.71 **(9.28)**	5.93 ± 2.14 **(15.37)**	5.28 ± 1.70 **(24.64)***	4.88 ± 1.15 **(30.31)***

TABLE 10

Leaf Length (cm) of *M. alba* Growing on Different Concentrations of Fly Ash Amended Soil During an Annual cycle of Three Consecutive Years

	Control	*5%*	*10%*	*20%*	*25%*	*30%*
March	6.78 ± 2.68	7.43 ± 1.68 **(9.58)**	8.72 ± 1.99 **(28.61)****	6.11 ± 2.79 **(9.89)**	5.43 ± 1.68 **(19.92)***	5.61 ± 2.74 **(17.26)***
April	7.32 ± 2.62	8.17 ± 2.97 **(11.61)**	9.91 ± 1.35 **(35.38)****	6.45 ± 1.42 **(11.89)**	5.68 ± 1.28 **(22.43)***	5.35 ± 1.46 **(26.91)***
May	7.82 ± 1.61	9.07 ± 2.91 **(15.98)***	10.12 ± 2.15 **(29.41)***	6.89 ± 2.45 **(11.89)**	5.92 ± 1.82 **(24.29)***	5.46 ± 1.74 **(30.17)****
June	8.13 ± 2.15	9.23 ± 3.16 **(13.53)**	10.17 ± 2.45 **(25.09)***	7.16 ± 1.26 **(11.93)**	6.11 ± 2.94 **(24.84)***	5.95 ± 1.87 **(26.81)****
July	8.81 ± 3.86	9.34 ± 3.56 **(6.01)**	11.67 ± 2.85 **(32.46)****	8.37 ± 3.34 **(4.99)**	6.76 ± 1.42 **(23.26)***	6.42 ± 1.67 **(27.12)***
August	9.32 ± 2.29	11.31 ± 3.72 **(16.85)***	13.53 ± 2.17 **(49.06)****	9.12 ± 1.27 **(3.34)**	7.08 ± 1.28 **(24.88)****	6.69 ± 1.29 **(26.32)***
September	9.49 ± 2.43	11.64 ± 2.39 **(23.06)***	14.99 ± 4.54 **(63.11)****	9.26 ± 3.09 **(1.41)**	7.07 ± 1.51 **(23.06)***	6.81 ± 1.24 **(25.89)****
October	8.62 ± 2.25	10.82 ± 2.04 **(26.82)****	12.92 ± 2.63 **(43.91)****	8.99 ± 3.16 **(2.14)**	6.81 ± 1.54 **(23.28)***	6.53 ± 2.95 **(26.18)****
November	8.35 ± 3.84	8.97 ± 3.16 **(9.29)**	11.05 ± 1.78 **(36.23)****	7.26 ± 2.87 **(8.89)**	5.84 ± 1.81 **(26.81)****	5.31 ± 1.69 **(36.71)****
December	6.25 ± 1.18	7.44 ± 1.49 **(19.04)***	7.53 ± 2.17 **(20.48)**	5.69 ± 1.97 **(8.96)**	5.28 ± 0.95 **(10.72)**	4.83 ± 1.42 **(22.72)***
January	5.49 ± 1.23	6.31 ± 1.15 **(14.93)**	6.96 ± 1.64 **(26.77)***	5.12 ± 1.79 **(6.73)**	4.93 ± 1.27 **(10.21)**	4.47 ± 1.54 **(18.57)***
February	6.44 ± 2.28	6.47 ± 1.76 **(0.48)**	8.25 ± 2.15 **(34.36)****	5.98 ± 2.84 **(2.60)**	5.32 ± 1.59 **(13.25)**	4.86 ± 2.38 **(20.84)***

TABLE 11

Leaf Length (cm) of *M. alba* Growing on Different Concentrations of Fly Ash Amended Soil During Different Stages of Growth

	Control	*5%*	*10%*	*20%*	*25%*	*30%*
Stage 1	7.51 ± 2.26	8.47 ± 2.68 **(12.81)**	9.73 ± 1.98 **(29.52)****	6.65 ± 1.98 **(11.45)**	5.78 ± 1.93 **(22.99)***	5.59 ± 1.95 **(25.56)****
Stage 2	9.10 ± 2.71	10.77 ± 2.92 **(18.35)***	13.40 ± 3.04 **(47.28)***	8.83 ± 2.69 **(2.94)**	6.95 ± 1.43 **(23.62)***	6.71 ± 1.78 **(26.37)****
Stage 3	6.56 ± 2.13	7.27 ± 1.89 **(10.73)**	8.54 ± 1.93 **(30.07)****	6.08 ± 2.36 **(7.30)**	5.49 ± 1.40 **(16.37)***	4.86 ± 1.75 **(25.88)****

TABLE 12

Leaf Width (cm) of *C. dichotoma* Growing on Different Concentrations of Fly Ash Amended Soil During an Annual Cycle of Three Consecutive Years

	Control	*5%*	*10%*	*20%*	*25%*	*30%*
March	5.27 ± 2.06	6.58 ± 1.61 **(24.85)***	5.85 ± 1.29 **(11.57)**	4.76 ± 2.35 **(9.67)**	4.51 ± 1.41 **(14.42)**	4.02 ± 0.93 **(23.72)***
April	5.33 ± 2.36	6.91 ± 2.27 **(29.64)****	5.24 ± 1.34 **(1.69)**	4.96 ± 1.27 **(6.94)**	4.64 ± 1.43 **(12.95)**	4.12 ± 1.07 **(22.71)***
May	5.73 ± 1.08	7.83 ± 2.31 **(36.65)****	5.66 ± 1.11 **(1.22)**	5.25 ± 1.23 **(8.37)**	4.91 ± 0.99 **(14.31)***	4.26 ± 1.44 **(25.65)****
June	5.81 ± 2.63	7.87 ± 1.19 **(35.46)****	5.74 ± 2.14 **(1.27)**	5.49 ± 1.09 **(5.54)**	4.92 ± 2.02 **(15.31)***	4.31 ± 1.16 **(25.84)***
July	6.65 ± 2.28	8.34 ± 1.31 **(25.41)***	6.26 ± 1.25 **(5.86)**	5.54 ± 2.17 **(16.69)***	5.25 ± 1.16 **(21.05)***	4.56 ± 1.42 **(31.42)***
August	6.91 ± 2.12	8.81 ± 2.13 **(27.49)****	6.66 ± 1.29 **(5.06)**	5.73 ± 1.16 **(17.07)***	5.32 ± 1.32 **(22.72)***	4.59 ± 1.27 **(33.57)***
September	7.08 ± 1.16	9.53 ± 2.23 **(34.65)****	6.69 ± 2.21 **(5.52)**	6.09 ± 1.15 **(13.98)**	5.74 ± 1.17 **(18.92)***	5.07 ± 1.04 **(28.38)****
October	7.27 ± 1.26	9.25 ± 1.33 **(27.23)****	7.17 ± 2.29 **(1.37)**	6.61 ± 1.21 **(9.07)**	5.96 ± 2.15 **(18.01)***	5.38 ± 0.94 **(25.99)***
November	6.56 ± 1.05	8.72 ± 2.49 **(32.92)****	5.98 ± 1.42 **(8.84)**	5.82 ± 1.41 **(11.28)**	4.72 ± 1.41 **(28.04)****	4.48 ± 1.32 **(31.77)***
December	5.16 ± 2.43	6.23 ± 1.45 **(20.73)***	5.04 ± 1.25 **(2..32)**	4.93 ± 1.29 **(4.45)**	4.69 ± 0.73 **(9.61)**	4.43 ± 0.72 **(14.34)**
January	5.03 ± 1.91	6.08 ± 1.15 **(20.82)***	4.98 ± 1.22 **(0.99)**	4.84 ± 1.28 **(3.77)**	4.63 ± 0.81 **(9.94)***	4.09 ± 1.13 **(18.68)***
February	5.82 ± 1.15	7.01 ± 0.56 **(20.44)***	5.44 ± 1.16 **(6.52)**	5.43 ± 0.41 **(6.71)**	5.13 ± 1.08 **(10.13)***	4.31 ± 1.17 **(25.94)***

TABLE 13

Leaf Width (cm) of *C. dichotoma* Growing on Different Concentrations of Fly Ash Amended Soil During Different Stages of Growth

	Control	*5%*	*10%*	*20%*	*25%*	*30%*
Stage 1	5.53 ± 2.03	7.29 ± 1.84 **(31.84)****	5.62 ± 1.47 **(1.58)**	5.11 ± 1.48 **(7.58)**	4.74 ± 1.46 **(14.27)**	4.17 ± 1.15 **(24.52)***
Stage 2	6.97 ± 1.69	8.98 ± 1.75 **(28.73)****	6.67 ± 1.76 **(4.40)**	5.99 ± 1.42 **(14.11)**	5.57 ± 1.42 **(20.13)***	4.90 ± 1.16 **(29.77)****
Stage 3	5.64 ± 1.63	7.01 ± 1.41 **(24.23)***	5.36 ± 1.18 **(5.01)**	5.25 ± 1.09 **(6.86)**	4.79 ± 0.95 **(15.06)***	4.32 ± 1.08 **(23.30)***

TABLE 14

Leaf Width (cm) of *M. alba* Growing on Different Concentrations of Fly Ash Amended Soil During an Annual Cycle of Three Consecutive Years

	Control	*5%*	*10%*	*20%*	*25%*	*30%*
March	5.05 ± 1.08	5.43 ± 1.47 **(7.52)**	5.47 ± 0.67 **(8.31)**	4.65 ± 1.07 **(7.92)**	4.43 ± 1.08 **(12.27)**	4.04 ± 1.11 **(20.15)***
April	5.88 ± 1.02	6.94 ± 1.33 **(18.02)***	7.17 ± 1.27 **(21.93)***	4.72 ± 1.03 **(19.72)***	4.58 ± 2.09 **(22.14)***	4.28 ± 1.10 **(28.91)***
May	5.96 ± 1.13	7.12 ± 2.52 **(19.46)***	7.78 ± 1.07 **(30.53)****	4.86 ± 2.11 **(18.62)***	4.26 ± 1.06 **(28.52)****	4.13 ± 1.09 **(30.71)****
June	6.22 ± 2.32	7.24 ± 2.35 **(16.39)***	8.34 ± 1.09 **(34.08)****	5.39 ± 1.24 **(13.34)**	4.34 ± 1.07 **(30.22)****	4.27 ± 1.14 **(31.35)****
July	6.25 ± 1.43	7.35 ± 2.32 **(20.48)***	8.59 ± 1.03 **(37.44)****	5.99 ± 1.08 **(4.16)**	4.85 ± 1.21 **(15.68)***	4.58 ± 1.03 **(22.88)***
August	7.08 ± 1.31	8.97 ± 2.49 **(27.03)***	9.93 ± 1.73 **(44.33)****	6.83 ± 2.07 **(12.06)**	5.27 ± 1.29 **(23.11)***	5.09 ± 1.08 **(26.07)****
September	7.89 ± 2.15	9.77 ± 1.05 **(23.82)***	10.62 ± 2.04 **(34.67)****	7.27 ± 2.07 **(7.85)**	5.29 ± 1.14 **(30.16)****	5.64 ± 1.19 **(33.84)****
October	7.44 ± 2.26	9.49 ± 1.05 **(27.55)****	9.68 ± 2.24 **(30.15)****	7.48 ± 2.08 **(0.53)**	5.16 ± 1.19 **(30.64)****	5.21 ± 1.25 **(33.66)****
November	6.38 ± 1.07	7.42 ± 2.34 **(16.55)***	8.18 ± 2.75 **(26.23)****	5.05 ± 1.36 **(22.06)***	5.03 ± 0.94 **(22.37)***	4.93 ± 1.28 **(23.91)***
December	4.24 ± 0.57	5.28 ± 1.22 **(24.52)***	5.92 ± 1.24 **(39.62)****	3.62 ± 0.94 **(8.25)**	3.29 ± 0.78 **(22.41)***	3.06 ± 1.46 **(27.83)****
January	4.14 ± 1.08	5.07 ± 0.65 **(22.46)***	6.67 ± 1.14 **(35.57)****	5.61 ± 1.05 **(12.31)**	3.47 ± 0.61 **(16.18)***	2.98 ± 1.08 **(28.01)****
February	4.65 ± 1.23	4.82 ± 1.06 **(3.65)**	6.07 ± 0.99 **(30.53)****	4.24 ± 0.85 **(8.81)**	3.83 ± 0.61 **(17.63)***	3.25 ± 1.28 **(30.14)***

TABLE 15

Leaf Width (cm) of *M. alba* Growing on Different Concentrations of Fly Ash Amended Soil During Different Stages of Growth

	Control	*5%*	*10%*	*20%*	*25%*	*30%*
Stage 1	5.77 ± 1.38	6.62 ± 1.91 **(15.66)***	7.19 ± 1.02 **(24.44)***	4.90 ± 1.36 **(15.14)**	4.40 ± 1.24 **(23.79)***	4.15 ± 1.01 **(28.08)***
Stage 2	7.11 ± 1.78	8.88 ± 1.77 **(24.84)***	9.70 ± 1.76 **(36.40)****	6.69 ± 1.81 **(5.87)**	5.31 ± 1.23 **(25.40)***	5.03 ± 1.12 **(29.23)****
Stage 3	4.87 ± 1.06	5.67 ± 1.35 **(16.40)***	6.64 ± 1.48 **(32.14)****	4.20 ± 1.05 **(13.84)**	3.90 ± 0.72 **(19.94)***	3.55 ± 0.70 **(27.11)****

TABLE 16

Total Chlorophyll (mg gm^{-1} fr. wt.) of the Leaves of *C. dichotoma* Growing on Different Concentrations of Fly Ash Amended Soil During an Annual Cycle of Three Consecutive Years

	Control	*5%*	*10%*	*20%*	*25%*	*30%*
March	1.25 ± 0.43	1.71 ± 0.21 **(40.92)****	0.89 ± 0.16 **(21.88)***	0.85 ± 0.07 **(35.28)****	0.81 ± 0.18 **(38.75)****	0.72 ± 0.09 **(42.24)****
April	1.47 ± 0.02	2.26 ± 0.04 **(53.74)****	1.45 ± 0.23 **(1.36)**	0.98 ± 0.05 **(33.03)****	0.93 ± 0.12 **(36.73)****	0.81 ± 0.06 **(44.89)****
May	1.43 ± 0.25	2.09 ± 0.15 **(46.15)****	1.29 ± 0.07 **(19.79)**	0.88 ± 0.04 **(38.46)****	0.85 ± 0.03 **(40.56)****	079 ± 0.03 **(44.75)****
June	1.32 ± 0.10	1.84 ± 0.17 **(39.12)****	1.19 ± 0.07 **(9.84)**	0.83 ± 0.04 **(37.12)****	0.76 ± 0.19 **(42.43)****	0.73 ± 0.03 **(44.69)****
July	1.19 ± 0.06	1.74 ± 0.03 **(47.06)****	0.94 ± 0.05 **(21.01)***	0.78 ± 0.16 **(34.45)****	0.76 ± 0.08 **(36.13)****	0.67 ± 0.11 **(43.69)****
August	1.08 ± 0.12	1.53 ± 0.08 **(41.66)****	0.86 ± 0.06 **(20.37)***	0.82 ± 0.03 **(24.07)***	0.78 ± 0.09 **(27.798)****	0.69 ± 0.03 **(36.11)****
September	0.94 ± 0.29	1.41 ± 0.17 **(50.86)****	0.85 ± 0.14 **(9.57)**	0.73 ± 0.03 **(22.34)***	0.69 ± 0.02 **(26.59)****	0.66 ± 0.17 **(29.78)****
October	0.89 ± 0.12	1.09 ± 0.04 **(22.47)***	0.81 ± 0.05 **(8.98)**	0.71 ± 0.14 **(20.22)***	0.55 ± 0.06 **(38.20)****	0.53 ± 0.08 **(40.49)****
November	0.75 ± 0.06	0.98 ± 0.18 **(30.67)****	0.81 ± 0.04 **(10.67)**	0.62 ± 0.04 **(17.33)***	0.62 ± 0.06 **(30.67)****	0.46 ± 0.11 **(38.20)****
December	0.72 ± 0.09	0.06 ± 0.25 **(47.22)****	0.69 ± 0.08 **(4.16)**	0.58 ± 0.05 **(19.41)***	0.46 ± 0.04 **(36.12)****	0.45 ± 0.03 **(37.51)****
January	0.68 ± 0.11	0.86 ± 0.17 **(26.47)****	0.71 ± 0.04 **(4.41)**	0.59 ± 0.07 **(13.23)**	0.53 ± 0.02 **(22.05)***	0.52 ± 0.05 **(23.53)***
February	0.71 ± 0.12	0.97 ± 0.06 **(36.62)****	0.67 ± 0.03 **(5.63)**	0.58 ± 0.07 **(18.31)***	0.48 ± 0.06 **(32.39)****	0.46 ± 0.02 **(35.21)****

TABLE 17

Total Chlorophyll (mg gm^{-1} fr. wt.) of the Leaves of *C. dichotoma* Growing on Different Concentrations of Fly Ash Amended Soil During Different Stages of Growth

	Control	*5%*	*10%*	*20%*	*25%*	*30%*
Stage 1	1.36 ± 0.21	1.97 ± 0.14 **(44.42)****	1.20 ± 0.13 **(11.88)**	0.88 ± 0.05 **(35.28)****	0.83 ± 0.11 **(38.76)****	0.76 ± 0.05 **(44.24)****
Stage 2	1.02 ± 0.14	1.44 ± 0.08 **(40.97)****	0.86 ± 0.07 **(15.61)***	0.76 ± 0.09 **(25.85)****	0.06 ± 0.62 **(32.19)****	0.63 ± 0.09 **(37.80)****
Stage 3	0.71 ± 0.09	0.96 ± 0.16 **(35.31)****	0.72 ± 0.04 **(1.39)**	0.59 ± 0.05 **(17.13)***	0.49 ± 0.04 **(30.42)****	0.47 ± 0.05 **(33.91)****

TABLE 18

Total Chlorophyll (mg gm⁻¹ fr. wt.) of the Leaves of *M. alba* Growing on Different Concentrations of Fly Ash Amended Soil During an Annual Cycle of Three Consecutive Years

	Control	*5%*	*10%*	*20%*	*25%*	*30%*
March	1.56 ± 0.09	1.78 ± 0.42 **(14.12)**	2.27 ± 0.65 **(45.51)****	1.47 ± 0.05 **(5.76)**	1.23 ± 0.32 **(21.15)***	1.09 ± 0.07 **(30.12)***
April	1.64 ± 0.41	2.64 ± 0.23 **(54.87)****	2.69 ± 0.32 **(64.02)****	1.65 ± 0.09 **(0.61)**	1.34 ± 0.08 **(18.29)***	1.23 ± 0.06 **(25.43)***
May	1.67 ± 0.03	2.29 ± 0.56 **(37.13)****	2.41 ± 0.26 **(44.31)****	1.58 ± 0.06 **(5.39)**	1.26 ± 0.32 **(24.56)***	1.07 ± 0.05 **(35.92)****
June	1.37 ± 0.08	1.67 ± 0.07 **(21.89)***	2.03 ± 0.08 **(48.17)****	1.21 ± 0.03 **(11.67)**	1.05 ± 0.07 **(22.63)***	0.93 ± 0.08 **(32.11)****
July	1.12 ± 0.32	1.68 ± 0.29 **(50.15)****	1.86 ± 0.53 **(66.07)****	0.95 ± 0.21 **(15.17)***	0.84 ± 0.08 **(25.52)****	0.69 ± 0.03 **(38.39)****
August	1.02 ± 0.08	1.64 ± 0.07 **(60.78)****	1.76 ± 0.24 **(72.54)***	1.09 ± 0.08 **(6.86)**	0.78 ± 0.06 **(28.43)***	0.84 ± 0.05 **(37.25)***
September	0.82 ± 0.04	1.18 ± 0.05 **(43.92)****	1.42 ± 0.05 **(73.17)****	0.75 ± 0.11 **(8.54)**	0.53 ± 0.08 **(35.36)****	0.57 ± 0.12 **(30.49)***
October	0.87 ± 0.07	1.22 ± 0.31 **(40.23)****	1.59 ± 0.09 **(82.76)****	0.79 ± 0.11 **(9.19)**	0.59 ± 0.08 **(62.18)****	0.57 ± 0.12 **(34.48)****
November	0.76 ± 0.07	0.94 ± 0.06 **(23.68)***	1.28 ± 0.23 **(68.42)****	0.78 ± 0.09 **(2.63)**	0.42 ± 0.06 **(14.73)****	0.48 ± 0.05 **(43.42)****
December	0.72 ± 0.08	0.91 ± 0.04 **(20.38)***	1.03 ± 0.42 **(43.06)****	0.48 ± 0.09 **(33.33)****	0.37 ± 0.06 **(48.61)****	0.35 ± 0.06 **(51.39)****
January	0.65 ± 0.09	0.88 ± 0.08 **(35.38)****	0.91 ± 0.07 **(40.13)****	0.59 ± 0.11 **(9.23)**	0.34 ± 0.07 **(47.69)****	0.33 ± 0.15 **(49.23)****
February	0.86 ± 0.07	1.21 ± 0.31 **(40.69)****	1.58 ± 0.09 **(83.72)****	0.79 ± 0.11 **(8.14)**	0.56 ± 0.08 **(34.88)****	0.51 ± 0.12 **(40.69)****

TABLE 19

Total Chlorophyll (mg gm^{-1} fr. wt.) of the Leaves of *M. alba* Growing on Different Concentrations of Fly Ash Amended Soil During Different Stages of Growth

	Control	*5%*	*10%*	*20%*	*25%*	*30%*
Stage 1	1.56 ± 0.15	2.07 ± 0.32 **(32.69)8***	2.35 ± 0.17 **(50.64)****	1.47 ± 0.06 **(5.28)**	1.22 ± 0.19 **(21.63)***	1.08 ± 0.06 **(30.77)****
Stage 2	0.95 ± 0.12	1.43 ± 0.18 **(49.34)****	1.65 ± 0.22 **(73.11)****	0.89 ± 0.12 **(6.53)**	0.67 ± 0.07 **(29.77)****	0.61 ± 0.08 **(35.51)****
Stage 3	0.74 ± 0.07	0.98 ± 0.12 **(31.17)****	1.20 ± 0.20 **(60.53)****	0.66 ± 0.11 **(11.70)**	0.42 ± 0.06 **(43.48)****	0.41 ± 0.09 **(45.81)***

TABLE 20

Chlorophyll a (mg gm^{-1} fr.wt.) of the Leaves of *C. dichotoma* Growing on Different Concentrations of Fly Ash Amended Soil During an Annual Cycle of Three Consecutive Years

	Control	*5%*	*10%*	*20%*	*25%*	*30%*
March	0.74 ± 0.17	0.87 ± 0.05 **(17.56)***	0.75 ± 0.19 **(1.35)**	0.71 ± 0.04 **(4.05)**	0.59 ± 0.06 **(20.27)***	0.52 ± 0.04 **(29.72)****
April	0.81 ± 0.13	0.98 ± 0.21 **(20.98)***	0.81 ± 0.07 **(0)**	0.75 ± 0.04 **(7.41)**	0.59 ± 0.06 **(27.16)***	0.54 ± 0.05 **(33.04)****
May	0.72 ± 0.05	0.89 ± 0.12 **(23.61)***	0.67 ± 0.05 **(6.94)**	0.63 ± 0.18 **(12.5)**	0.54 ± 0.05 **(25.43)***	0.49 ± 0.07 **(31.94)****
June	0.72 ± 0.08	0.87 ± 0.05 **(20.83)***	0.64 ± 0.04 **(11.11)**	0.53 ± 0.06 **(26.39)****	0.42 ± 0.07 **(41.67)****	0.38 ± 0.04 **(47.22)****
July	0.75 ± 0.11	0.84 ± 0.09 **(12.53)***	0.61 ± 0.03 **(18.67)***	0.51 ± 0.04 **(32.73)****	0.49 ± 0.05 **(34.67)****	0.41 ± 0.09 **(45.33)****
August	0.67 ± 0.04	0.82 ± 0.05 **(22.38)***	0.63 ± 0.13 **(5.97)**	0.48 ± 0.04 **(28.35)****	0.42 ± 0.05 **(37.31)****	0.35 ± 0.07 **(47.76)****
September	0.63 ± 0.12	0.81 ± 0.06 **(28.57)****	0.57 ± 0.07 **(9.52)**	0.48 ± 0.06 **(23.81)***	0.37 ± 0.04 **(41.26)****	0.33 ± 0.05 **(47.62)****
October	0.55 ± 0.13	0.71 ± 0.15 **(29.09)****	0.52 ± 0.02 **(5.45)**	0.41 ± 0.05 **(25.45)***	0.34 ± 0.06 **(38.18)****	0.31 ± 0.04 **(43.64)****
November	0.51 ± 0.060.04	0.67 ± 0.02 **(31.37)****	0.50 ± 0.05 **(1.96)**	0.46 ± 0.03 **(9.81)**	0.31 ± 0.05 **(39.22)****	0.29 ± **(43.14)****
December	0.48 ± 0.07	0.52 ± 0.03 **(8.33)***	0.42 ± 0.03 **(12.05)**	0.39 ± 0.06 **(18.75)***	0.36 ± 0.04 **(25.72)****	0.29 ± 0.05 **(46.51)****
January	0.43 ± 0.02	0.51 ± 0.07 **(18.64)***	0.39 ± 0.04 **(9.32)**	0.29 ± 0.04 **(32.55)***	0.26 ± 0.05 **(39.53)****	0.23 ± 0.06 **(46.51)****
February	0.45 ± 0.02	0.62 ± 0.14 **(37.78)****	0.31 ± 0.04 **(31.11)****	0.32 ± 0.03 **(28.89)****	0.27 ± 0.04 **(40.62)****	0.28 ± 0.05 **(37.78)****

TABLE 21

Chorophyll a (mg gm^{-1} fr. wt.) of the Leaves of *C. dichotoma* Growing on Different Concentrations of Fly Ash Amended Soil During Different Stages of Growth

	Control	*5%*	*10%*	*20%*	*25%*	*30%*
Stage 1	0.74 ± 0.10	0.90 ± 0.10 **(20.73)***	0.71 ± 0.08 **(4.01)**	0.65 ± 0.08 **(12.37)**	0.53 ± 0.06 **(28.43)****	0.48 ± 0.05 **(35.45)****
Stage 2	0.65 ± 0.11	0.79 ± 0.08 **(20.73)***	0.58 ± 0.06 **(4.01)**	0.47 ± 0.04 **(12.37)**	0.40 ± 0.05 **(28.43)****	0.35 ± 0.06 **(35.45)****
Stage 3	0.46 ± 0.04	0.58 ± 0.07 **(24.06)***	0.40 ± 0.03 **(13.37)**	0.36 ± 0.04 **(21.92)***	0.32 ± 0.04 **(35.83)****	0.27 ± 0.05 **(41.71)****

TABLE 22

Chlorophyll a (mg gm^{-1} fr. wt.) of the Leaves of *M. alba* Growing on Different Concentrations of Fly Ash Amended Soil During an Annual Cycle of Three Consecutive Years

	Control	*5%*	*10%*	*20%*	*25%*	*30%*
March	1.07 ± 0.04	1.62 ± 0.03 **(51.41)****	1.69 ± 0.05 **(57.94)***	0.96 ± 0.03 **(10.28)**	0.84 ± 0.19 **(21.49)***	0.83 ± 0.02 **(22.42)***
April	1.17 ± 0.09	1.88 ± 0.09 **(60.68)****	1.89 ± 0.21 **(61.53)****	1.05 ± 0.17 **(10.25)**	0.94 ± 0.07 **(19.65)***	0.86 ± 0.07 **(26.49)****
May	1.14 ± 0.09	1.83 ± 0.27 **(60.53)****	1.87 ± 0.09 **(64.03)***	1.01 ± 0.05 **(11.41)**	0.85 ± 0.06 **(25.43)***	0.83 ± 0.06 **(27.19)****
June	0.99 ± 0.04	1.57 ± 0.14 **(58.59)****	1.65 ± 0.26 **(66.69)***	0.91 ± 0.07 **(8.08)**	0.71 ± 0.14 **(28.13)****	0.77 ± 0.05 **(22.52)***
July	0.87 ± 0.04	1.11 ± 0.02 **(27.58)****	1.32 ± 0.14 **(51.72)***	0.81 ± 0.19 **(6.89)**	0.65 ± 0.04 **(25.28)****	0.62 ± 0.03 **(28.73)****
August	0.94 ± 0.14	1.32 ± 0.27 **(40.43)****	1.45 ± 0.08 **(54.26)****	0.89 ± 0.03 **(5.32)**	0.69 ± 0.04 **(26.59)****	0.62 ± 0.05 **(34.04)****
September	0.81 ± 0.12	1.46 ± 0.23 **(80.24)****	1.57 ± 0.21 **(93.82)****	0.96 ± 0.18 **(18.51)***	0.75 ± 0.05 **(7.41)**	0.72 ± 0.13 **(11.87)**
October	0.65 ± 0.04	0.89 ± 0.14 **(36.92)****	1.03 ± 0.15 **(56.92)****	0.63 ± 0.04 **(3.07)**	0.49 ± 0.02 **(24.62)***	0.36 ± 0.03 **(44.65)****
November	0.55 ± 0.02	0.75 ± 0.04 **(36.48)****	0.89 ± 0.03 **(61.81)***	0.47 ± 0.05 **(14.54)**	0.45 ± 0.03 **(18.82)***	0.31 ± 0.02 **(43.64)****
December	0.43 ± 0.03	0.63 ± 0.18 **(46.52)****	0.69 ± 0.05 **(60.46)****	0.42 ± 0.04 **(2.33)**	0.32 ± 0.07 **(25.58)****	0.28 ± 0.05 **(34.89)****
January	0.46 ± 0.02	0.46 ± 0.08 **(39.13)****	0.76 ± 0.09 **(65.21)****	0.53 ± 0.03 **(15.21)***	0.31 ± 0.04 **(32.68)*****	0.26 ± 0.04 **(43.47)****
February	0.51 ± 0.05	0.72 ± 0.15 **(41.18)****	0.75 ± 0.02 **(47.06)****	0.48 ± 0.03 **(5.89)**	0.35 ± 0.04 **(31.37)****	0.33 ± 0.03 **(35.29)****

TABLE 23

Chlorophyll a (mg gm^{-1} fr. wt.) of the Leaves of *M. alba* Growing on Different Concentrations of Fly Ash Amended Soil During Different Stages of Growth

	Control	*5%*	*10%*	*20%*	*25%*	*30%*
Stage 1	1.09 ± 0.06	1.72 ± 0.13 **(57.89)****	1.77 ± 0.15 **(62.47)****	0.98 ± 0.08 **(10.07)**	0.83 ± 0.11 **(23.57)***	0.82 ± 0.05 **(24.71)***
Stage 2	0.81 ± 0.08	1.19 ± 0.16 **(46.17)****	1.34 ± 0.13 **(63.91)****	0.82 ± 0.11 **(0.61)**	0.64 ± 0.03 **(21.10)****	0.58 ± 0.06 **(29.05)****
Stage 3	0.48 ± 0.03	0.68 ± 0.11 **(40.51)****	0.77 ± 0.14 **(58.46)****	0.47 ± 0.03 **(2.56)**	0.35 ± 0.04 **(26.67)****	0.29 ± 0.03 **(39.48)****

TABLE 24

Chlorphyll b (mg gm^{-1} fr. wt.) of Leaves of *C. dichotoma* growing on different concentration of fly ash amended soil during an annual cycle of three consecutive years.

	Control	*5%*	*10%*	*20%*	*25%*	*30%*
March	0.51 ± 0.09	0.65 ± 0.03 **(27.45)****	0.48 ± 0.04 **(5.88)**	0.45 ± 0.04 **(11.76)**	0.43 ± 0.05 **(15.68)***	0.38 ± 0.03 **(25.49)****
April	0.58 ± 0.03	0.71 ± 0.04 **(22.41)***	0.52 ± 0.04 **(10.34)**	0.45 ± 0.05 **(22.41)***	0.36 ± 0.06 **(37.93)****	0.34 ± 0.03 **(41.37)****
May	0.56 ± 0.05	0.68 ± 0.04 **(21.42)***	0.53 ± 0.03 **(5.35)**	0.47 ± 0.03 **(23.21)***	0.41 ± 0.03 **(26.78)****	0.36 ± 0.05 **(35.71)****
June	0.47 ± 0.07	0.63 ± 0.03 (31.91)**	0.43 ± 0.05 (8.51)	0.39 ± 0.04 (17.02)**	0.35 ± 0.04 (25.53)*8	0.33 ± 0.05 (29.78)**
July	0.43 ± 0.04	0.59 ± 0.09 **(37.31)****	0.37 ± 0.03 **(13.95)**	0.33 ± 0.02 **(23.26)****	0.31 ± 0.03 **(27.91)****	0.28 ± 0.04 **(34.88)****
August	0.42 ± 0.03	0.55 ± 0.04 **(30.95)****	0.39 ± 0.05 **(7.14)**	0.37 ± 0.03 **(11.91)**	0.29 ± 0.04 **(30.85)****	0.26 ± 0.06 **(38.09)****
September	0.43 ± 0.04	0.52 ± 0.05 **(20.93)***	0.37 ± 0.08 **(13.95)**	0.36 ± 0.02 **(16.27**	0.31 ± 0.04 **(27.91)****	0.28 ± 0.05 **(34.88)****
October	0.41 ± 0.03	0.53 ± 0.04 **(29.26)****	0.37 ± 0.02 **(9.76)**	0.33 ± 0.04 **(19.51)***	0.28 ± 0.05 **(31.71)****	0.25 ± 0.03 **(39.02)****
November	0.37 ± 0.02	0.47 ± 0.03 **(27.03)****	0.33 ± 0.07 **(10.81)**	0.29 ± 0.03 **(21.62)***	0.24 ± 0.03 **(35.14)****	0.25 ± 0.05 **(32.43)****
December	0.34 ± 0.08	0.39 ± 0.07 **(14.71)**	0.28 ± 0.07 **(17.65)***	0.27 ± 0.04 **(20.59)***	0.21 ± 0.05 **(38.24)****	0.21 ± 0.03 **(38.24)****
January	0.32 ± 0.05	0.38 ± 0.02 **(18.63)***	0.27 ± 0.02 **(15.83)***	0.25 ± 0.04 **(20.63)***	0.19 ± 0.06 **(40.85)****	0.18 ± 0.02 **(43.75)****
February	0.32 ± 0.03	0.42 ± 0.04 **(31.25)****	0.31 ± 0.02 **(3.13)**	0.29 ± 0.04 **(9.38)**	0.21 ± 0.02 **(34.38)****	0.19 ± 0.03 **(40.63)****

TABLE 25

Chlorophyll b (mg gm^{-1} fr. wt.) of the Leaves of *C. dichotoma* Growing on Different Concentrations of Fly Ash Amended Soil During Different Stages of Growth

	Control	*5%*	*10%*	*20%*	*25%*	*30%*
Stage 1	0.53 ± 0.06	0.66 ± 0.03 **(25.47)***	0.49 ± 0.04 **(7.55)**	0.43 ± 0.04 **(18..87)***	0.38 ± 0.04 **(26.88)****	0.35 ± 0.04 **(33.49)****
Stage 2	0.42 ± 0.03	0.54 ± 0.05 **(29.58)****	0.37 ± 0.04 **(11.24)**	0.34 ± 0.02 **(17.75)***	0.29 ± 0.04 **(29.58)****	0.26 ± 0.04 **(36.68)****
Stage 3	0.33 ± 0.04	0.41 ± 0.04 **(22.96)***	0.29 ± 0.04 **(11.85)**	0.27 ± 0.03 **(18.51)***	0.21 ± 0.04 **(37.04)****	0.20 ± 0.03 **(38.52)****

TABLE 26

Chlorophyll b (mg gm⁻¹ fr. wt.) of the Leaves of *M. alba* Growing on Different Concentration of Fly Ash Amended Soil During an Annual Cycle of Three Consecutive Years.

	Control	*5%*	*10%*	*20%*	*25%*	*30%*
March	0.58 ± 0.66	0.70 ± 0.08 **(36.21)****	0.85 ± 0.29 **(46.55)****	0.61 ± 0.05 **(5.17)**	0.46 ± 0.06 **(20.69)***	0.43 ± 0.04 **(25.86)****
April	0.63 ± 0.04	0.95 ± 0.07 **(50.79)****	0.97 ± 0.15 **(53.96)****	0.76 ± 0.13 **(20.64)***	0.62 ± 0.05 **(1.69)**	0.58 ± 0.04 **(7.93)**
May	0.66 ± 0.17	0.91 ± 0.28 **(37.87)***	0.95 ± 0.09 **(43.93)****	0.64 ± 0.02 **(3.4)***	0.51 ± 0.06 **(22.72)**	0.45 ± 0.05 **(31.82)****
June	0.59 ± 0.02	0.83 ± 0.14 **(40.67)****	0.89 ± 0.07 **(50.84)****	0.58 ± 0.16 **(1.69)**	0.43 ± 0.09 **(27.11)****	0.44 ± 0.08 **(25.42)***
July	0.52 ± 0.07	0.74 ± 0.06 **(42.31)****	0.82 ± 0.16 **(57.69)****	0.53 ± 0.07 **(1.92)**	0.39 ± 0.03 **(25.38)**	0.36 ± 0.02 **(30.76)****
August	0.46 ± 0.07	0.71 ± 0.18 **(54.35)****	0.72 ± 0.14 **(56.52)****	0.45 ± 0.06 **(2.17)**	0.32 ± 0.08 **(30.43)****	0.31 ± 0.05 **(32.61)****
September	0.43 ± 0.03	0.63 ± 0.14 **(46.51)****	0.67 ± 0.09 **(55.82)****	0.41 ± 0.02 **(4.65)**	0.31 ± 0.08 **(27.91)****	0.27 ± 0.01 **(37.29)****
October	0.38 ± 0.06	0.59 ± 0.08 **(55.26)****	0.61 ± 0.05 **(60.52)****	0.33 ± 0.02 **(13.16)**	0.25 ± 0.04 **(31.21)****	0.24 ± 0.02 **(36.84)****
November	0.32 ± 0.03	0.55 ± 0.07 **(71.87)****	0.62 ± 0.08 **(93.75)****	0.31 ± 0.05 **(3.12)**	0.22 ± 0.02 **(31.26)****	0.21 ± 0.04 **(34.38)****
December	0.29 ± 0.02	0.49 ± 0.05 **(68.96)****	0.50 ± 0.06 **(72.41)***	0.22 ± 0.09 **(24.13)***	0.21 ± 0.02 **(27.56)****	0.18 ± 0.04 **(37.93)****
January	0.28 ± 0.05	0.48 ± 0.04 **(71.42)****	0.51 ± 0.07 **(82.14)***	0.26 ± 0.02 **(7.14)**	0.19 ± 0.06 **(31.18)****	0.15 ± 0.03 **(46.42)****
February	0.35 ± 0.05	0.54 ± 0.08 **(54.29)****	0.59 ± 0.04 **(68.58)****	0.35 ± 0.07 **(0)**	0.21 ± 0.01 **(40)****	0.19 ± 0.02 **(45.71)****

TABLE 27

Chlorophyll b (mg gm^{-1} fr. wt.) of the Leaves of *M. alba* Growing on Different Concentrations of Fly Ash Amended Soil During Different Stages of Growth

	Control	*5%*	*10%*	*20%*	*25%*	*30%*
Stage 1	0.61 ± 0.07	0.87 ± 0.14 **(41.46)****	0.91 ± 0.15 **(48.78)***	0.64 ± 0.09 **(5.28)**	0.51 ± 0.06 **(17.88)***	0.47 ± 0.05 **(22.76)***
Stage 2	0.44 ± 0.05	0.66 ± 0.06 **(39.16)****	0.71 ± 0.11 **(57.54)***	0.43 ± 0.04 **(3.91)**	0.31 ± 0.05 **(29.05)****	0.29 ± 0.04 **(34.70)****
Stage 3	0.31 ± 0.03	0.51 ± 0.06 **(66.13)****	0.55 ± 0.06 **(79.03)***	0.28 ± 0.05 **(8.06)**	0.21 ± 0.02 **(33.06)****	0.18 ± 0.03 **(41.13)****

TABLE 28

Carotenoids (mg gm⁻¹ fr. wt.) of the Leaves of *C. dichotoma* Growing on Different Concentration of Fly Ash Amended Soil During an Annual Cycle of Three Consecutive Years.

	Control	*5%*	*10%*	*20%*	*25%*	*30%*
March	0.56 ± 0.08	0.63 ± 0.04 **(12.57)**	0.56 ± 0.04 **(0)**	0.48 ± 0.06 **(14.29)**	0.45 ± 0.03 **(19.64)***	0.39 ± 0.07 **(30.35)****
April	0.59 ± 0.04	0.71 ± 0.02 **(20.34)***	0.46 ± 0.09 **(22.03)***	0.43 ± 0.03 **(27.12)****	0.41 ± 0.06 **(30.51)****	0.37 ± 0.04 **(37.29)****
May	0.59 ± 0.04	0.79 ± 0.15 **(33.89)****	0.47 ± 0.09 **(20.34)****	0.46 ± 0.03 **(22.03)***	0.43 ± 0.06 **(27.12)**	0.38 ± 0.04 **(35.59)****
June	0.54 ± 0.02	0.63 ± 0.07 **(17.67)***	0.51 ± 0.06 **(5.56)**	0.43 ± 0.04 **(20.37)***	0.41 ± 0.06 **(24.07)***	0.36 ± 0.02 **(33.33)****
July	0.48 ± 0.03	0.59 ± 0.04 **(22.92)***	0.51 ± 0.02 **(6.25)****	0.38 ± 0.05 **(20.83)***	0.35 ± 0.06 **(27.08)****	0.33 ± 0.05 **(31.25)****
August	0.42 ± 0.02	0.58 ± 0.03 **(38.09)****	0.42 ± 0.05 **(0)**	0.36 ± 0.04 **(14.29)**	0.33 ± 0.06 **(21.42)***	0.32 ± 0.07 **(23.81)***
September	0.45 ± 0.03	0.54 ± 0.07 **(20.23)***	0.39 ± 0.04 **(13.33)**	0.32 ± 0.03 **(28.89)****	0.31 ± 0.05 **(31.11)****	0.28 ± 0.06 **(37.78)****
October	0.38 ± 0.03	0.49 ± 0.03 **(28.94)****	0.35 ± 0.04 **(7.89)****	0.31 ± 0.05 **(18.42)**	0.33 ± 0.06 **(13.16)**	0.27 ± 0.05 **(28.95)****
November	0.36 ± 0.04	0.43 ± 0.06 **(19.44)***	0.32 ± 0.02 **(11.03)**	03531 ± 0.03 **(2.78)**	0.28 ± 0.03 **(22.12)***	0.25 ± 0.05 **(30.56)****
December	0.32 ± 0.04	0.41 ± 0.07 **(28.13)***	0.29 ± 0.02 **(9.38)**	0.28 ± 0.07 **(12.50)**	0.25 ± 0.02 **(21.88)***	0.23 ± 0.04 **(2813)****
January	0.29 ± 0.05	0.36 ± 0.02 **(24.14)***	0.27 ± 0.03 **(8.89)**	0.21 ± 0.03 **(27.69)****	0.18 ± 0.04 **(37.93)****	0.17 ± 0.02 **(41.37)****
February	0.31 ± 0.03	0.42 ± 0.08 **(35.48)****	0.28 ± 0.04 **(9.67)**	0.25 ± 0.05 **(19.35)***	0.21 ± 0.05 **(32.26)****	0.19 ± 0.04 **(38.71)****

TABLE 29

Carotenoids (mg gm^{-1} fr. wt.) of the Leaves of *C. dichotoma* Growing on Different Concentrations of Fly Ash Amended Soil During Different Stages of Growth

	Control	*5%*	*10%*	*20%*	*25%*	*30%*
Stage 1	0.57 ± 0.04	0.69 ± 0.07 **(21.05)***	0.5 ± 0.07 **(12.28)**	0.45 ± 0.04 **(21.05)***	0.42 ± 0.04 **(25.44)***	0.37 ± 0.04 **(34.21)****
Stage 2	0.43 ± 0.02	0.55 ± 0.04 **(27.17)****	0.41 ± 0.03 **(3.47)**	0.34 ± 0.04 **(20.81)***	0.33 ± 0.05 **(23.69)***	0.3 ± 0.05 **(30.63)****
Stage 3	0.32 ± 0.04	0.40 ± 0.05 **(26.56)****	0.29 ± 0.03 **(9.37)**	0.27 ± 0.04 **(14.84)**	0.23 ± 0.02 **(28.12)****	0.21 ± 0.03 **(34.37)****

TABLE 30

Carotenoids (mg gm^{-1} fr. wt.) of the Leaves of *C. alba* Growing on Different Concentration of Fly Ash Amended Soil During an annual Cycle of Three Consecutive Years.

	Control	*5%*	*10%*	*20%*	*25%*	*30%*
March	0.73 ± 0.04	0.88 ± 0.26 **(2054)***	0.91 ± 0.15 **(24.65)***	0.74 ± 0.07 **(1.37)**	0.61 ± 0.19 **(16.43)***	0.59 ± 0.11 **(19.17)***
April	0.79 ± 0.22	0.92 ± 0.06 **(16.46)***	0.06 ± 0.13 **(34.18)****	0.74 ± 0.08 **(6.32)****	0.62 ± 0.05 **(21.52)***	0.57 ± 0.08 **(27.84)****
May	0.72 ± 0.03	0.84 ± 0.14 **(16.67)***	0.96 ± 0.08 **(33.06)****	0.71 ± 0.16 **(1.39)**	0.56 ± 0.08 **(22.14)***	0.52 ± 0.02 **(27.18)****
June	0.69 ± 0.19	0.83 ± 0.06 **(20.29)***	0.98 ± 0.19 **(42.03)****	0.64 ± 0.12 **(7.25)**	0.53 ± 0.07 **(23.19)***	0.49 ± 0.09 **(28.98)****
July	0.65 ± 0.04	0.81 ± 0.08 **(24.62)***	0.92 ± 0.07 **(41.53)****	0.65 ± 0.09 **(0)**	0.57 ± 0.07 **(12.37)**	0.38 ± 0.03 **(41.53)****
August	0.56 ± 0.16	0.78 ± 0.07 **(39.28)****	0.82 ± 0.09 **(46.42)****	0.54 ± 0.11 **(3.57)**	0.42 ± 0.07 **(25.69)****	0.31 ± 0.04 **(44.58)****
September	0.59 ± 0.08	0.82 ± 0.14 **(38.98)****	0.89 ± 0.16 **(50.43)****	0.57 ± 0.09 **(3.38)**	0.47 ± 0.05 **(20.14)****	0.38 ± 0.07 **(35.28)****
October	0.49 ± 0.09	0.71 ± 0.06 **(44.93)****	0.81 ± 0.15 **(60.85)****	0.48 ± 0.05 **(2.09)**	0.36 ± 0.04 **(26.34)****	0.25 ± 0.04 **(48.59)****
November	0.41 ± 0.02	0.64 ± 0.09 **(56.09)****	0.69 ± 0.05 **(68.29)**	0.38 ± 0.02 **(7.31)**	0.31 ± 0.05 **(24.39)***	0.26 ± 0.03 **(36.58)****
December	0.39 ± 0.04	0.57 ± 0.08 **(40.15)****	0.62 ± 0.05 **(58.97)****	0.37 ± 0.05 **(5.12)**	0.25 ± 0.6 **(35.89)****	0.23 ± 0.02 **(41.02)****
January	0.38 ± 0.02	0.51 ± 0.03 **(34.21)****	0.54 ± 0.02 **(42.15)****	0.34 ± 0.07 **(10.52)**	0.23 ± 0.02 **(39.47)****	0.19 ± 0.03 **(50.36)****
February	0.42 ± 0.03	0.62 ± 0.09 **(47.61)****	0.63 ± 0.06 **(50.93)**	0.41 ± 0.05 **(2.38)**	0.32 ± 0.02 **(23.81)***	0.29 ± 0.02 **(30.95)**

TABLE 31

Carotenoids (mg gm^{-1} fr. wt.) of the Leaves of *M. alba* Growing on Different Concentrations of Fly Ash Amended Soil During Different Stages of Growth

	Control	*5%*	*10%*	*20%*	*25%*	*30%*
Stage 1	0.73 ± 0.12	0.86 ± 0.13 **(18.43)***	0.97 ± 0.13 **(33.45)****	0.71 ± 0.11 **(3.41)**	0.58 ± 0.09 **(20.82)***	0.54 ± 0.07 **(25.94)****
Stage 2	0.57 ± 0.09	0.78 ± 0.08 **(36.24)****	0.86 ± 0.11 **(50.22)8**	0.56 ± 0.08 **(2.18)**	0.45 ± 0.05 **(20.52)***	0.33 ± 0.04 **(42.36)****
Stage 3	0.42 ± 0.02	0.58 ± 0.07 **(46.25)****	0.62 ± 0.04 **(55)****	0.37 ± 0.04 **(6.25)**	0.37 ± 0.03 **(30.63)****	0.24 ± 0.02 **(39.37)****

TABLE 32

Sugar Content (μ gm g[-1] fr.wt.) of the Leaves of *C. dichotoma* Growing on Different Concentrations of Fly Ash Amended Soil During an Annual Cycle of Three Consecutive Years

	Control	*5%*	*10%*	*20%*	*25%*	*30%*
March	1321.64 ± 42.48	1916.25 ± 36.67 **(44.91)***	1206.86 ± 44.61 **(8.84)**	1082.47 ± 24.71 **(18.09)***	955.25 ± 28.25 **(27.72)****	998.83 ± 38.09 **(28.21)****
April	1514.61 ± 32.48	2325.37 ± 67.93 **(53.52)****	1434.27 ± 83.36 **(5.31)**	1275.17 ± 50.89 **(15.81)***	1162.68 ± 24.11 **(27.54)****	1092.45 ± 22.72 **(24.26)***
May	1196.53` ± 36.04	2055.63 ± 88.43 **(27.15)****	1089.14 ± 46.68 **(8.99)**	948.52 ± 55.81 **(20.74)***	867.22 ± 39.08 **(27.54)****	906.18 ± 56.48 **(24.26)***
June	1432.53 ± 84.23	2055.63 ± 132.21 **(43.49)****	1281.46 ± 69.53 **(10.54)**	1135.05 ± 46.73 **(20.76)**	1021.37 ± 72.13 **(28.70)****	997.18 ± 20.53 **(30.39)****
July	3861.28 ± 29.81	962.32 ± 16.96 **(11.73)**	737.03 ± 15.31 **(14.42)**	669.85 ± 11.38 **(22.23)***	628.94 ± 14.63 **(26.98)****	563.48 ± 10.91 **(34.57)****
August	1340.87 ± 43.31	1544.85 ± 22.91 **(15.21)***	1299.21 ± 36.32 **(3.11)**	1190.47 ± 28.89 **(11.22)**	1014.45 ± 26.18 **(24.34)***	999.24 ± 29.67 **(25.48)***
September	1184.62 ± 34.83	1424.46 ± 41.06 **(20.25)***	1096.25 ± 42.91 **(7.46)**	952.56 ± 28.57 **(19.59)**	840.92 ± 40.13 **(29.01)****	716.13 ± 16.29 **(39.54)****
October	1272.63 ± 34.83	1710.45 ± 41.06 **(20.25)***	1108.57 ± 42.91 **(7.46)**	970.44 ± 38.57 **(19.59)***	853.21 ± 40.13 **(29.01)****	892.13 ± 16.29 **(25.54)****
November	782.43 ± 14.85	984.87 ± 26.31 **(25.87)****	668.82 ± 20.48 **(14.52)**	564.36 ± 36.81 **(27.87)****	462.45 ± 14.24 **(40.89)****	445.63 ± 20.01 **(43.05)***
December	496.85 ± 26.03	634.91 ± 38.46 **(27.78)****	395.42 ± 29.81 **(20.41)***	386.71 ± 14.61 **(22.17)****	353.62 ± 16.72 **(28.82)****	280.51 ± 29.48 **(43.55)****
January	1072.52 ± 52.21	1368.21 ± 45.32 **(27.56)****	1008.46 ± 32.28 **(5.97)**	852.33 ± 27.96 **(20.53)***	736.18 ± 44.25 **(31.36)****	688.51 ± 24.17 **(35.81)****
February	844.37 ± 18.83	1083.42 ± 26.59 **(28.31)****	878.49 ± 18.95 **(4.05)**	734.45 ± 27.44 **(13.02)**	690.86 ± 26.43 **(18.18)***	587.85 ± 13.56 **(30.38)****

TABLE 33

Sugar Contents (µ gm g^{-1} fr. wt.) of the Leaves of *C. dichotoma* Growing on Different Concentrations of Fly Ash Amended Soil During Different Stages of Growth

	Control	*5%*	*10%*	*20%*	*25%*	*30%*
Stage 1	1366.40 ± 48.80	1954.76 ± 81.31 **(43.05)****	1252.43 ± 61.04 **(8.34)**	1110.30 ± 44.53 **(18.74)***	1001.63 ± 40.89 **(26.69)***	966.22 ± 34.45 **(27.82)****
Stage 2	1164.84 ± 33.28	1410.52 ± 30.25 **(21.09)***	1060.26 ± 32.98 **(0.98)**	945.83 ± 24.25 **(18.80)***	834.38 ± 26.71 **(28.37)****	792.76 ± 19.03 **(31.94)****
Stage 3	799.04 ± 27.98	1017.85 ± 34.17 **(27.38)****	737.79 ± 25.38 **(7.67)**	634.46 ± 26.51 **(20.59)***	560.78 ± 25.41 **(29.82)****	500.61 ± 21.81 **(37.35)***

TABLE 34

Sugar Content (μ gm g⁻¹ fr.wt.) of the Leaves of *C. alba* Growing on Different Concentrations of Fly Ash Amended Soil During an Annual Cycle of Three Consecutive Years

	Control	*5%*	*10%*	*20%*	*25%*	*30%*
March	1691.27 ± 16.87	2014.46 ± 4226 **(19.11)***	2459.27 ± 73.75 **(42.41)***	1384.95 ± 48.57 **(18.12)***	1361.86 ± 24.07 **(19.47)***	1007.98 ± 28.38 **(40.14)&***
April	2032.54 ± 35.96	2251.65 ± 24.08 **(10.78)**	2678.92 ± 47.14 **(31.81)****	1956.23 ± 30.17 **(3.75)**	1524.74 ± 39.75 **(24.98)***	1433.75 ± 18.64 **(29.46)***
May	2391.54 ± 68.54	2603.16 ± 77.24 **(8.84)**	2935.87 ± 54.32 **(22.76)***	2367.69 ± 23.35 **(0.99)**	1840.35 ± 40.08 **(23.04)***	1629.48 ± 32.22 **(31.86)****
June	1385.64 ± 41.32	1524.83 ± 24.12 **(10.05)**	1839.25 ± 36.47 **(32.74)****	1272.85 ± 40.13 **(8.14)**	968.49 ± 17.89 **(30.15)****	903.62 ± 26.56 **(34.78)****
July	568.43 ± 28.09	705.62 ± 15.79 **(24.14)***	856.86 ± 38.21 **(50.74)****	484.81 ± 16.39 **(14.71)**	341.65 ± 24.54 **(39.89)****	327.66 ± 14.71 **(42.36)****
August	1811.27 ± 39.53	2316.43 ± 57.89 **(27.89)****	2550.43 ± 27.64 **(40.81)***	1785.63 ± 34.61 **(1.42)**	1468.85 ± 46.59 **(18.94)***	1252.35 ± 23.81 **(30.85)****
September	1451.24 ± 29.16	1603.27 ± 14.23 **(10.47)**	1992.29 ± 72.28 **(37.28)***	1502.45 ± 57.93 **(37.28)****	1085.66 ± 46.14 **(25.19)***	915.28 ± 20.49 **(36.93)****
October	1278.45 ± 72.21	1544.48 ± 37.44 **(20.88)***	1745.67 ± 27.18 **(36.54)****	1158.68 ± 40.49 **(9.36)**	844.34 ± 36.53 **(33.95)****	718.53 ± 26.32 **(43.79)****
November	765.27 ± 31.84	916.45 ± 44.46 **(19.76)***	1235.73 ± 67.17 **(61.47)****	634.11 ± 25.61 **(17.13)***	656.87 ± 53.75 **(14.06)**	484.51 ± 38.52 **(36.68)****
December	553.06 ± 12.45	687.27 ± 29.61 **(24.27)***	807.93 ± 12.48 **(46.08)****	481.62 ± 15.29 **(12.91)**	376.55 ± 26.41 **(31.92)****	312.83 ± 19.04 **(43.65)****
January	685.69 ± 23.74	824.82 ± 42.94 **(20.29)***	899.25 ± 25.98 **(31.41)****	772.83 ± 13.56 **(12.78)**	468.43 ± 26.72 **(31.64)****	403.69 ± 28.95 **(41.13)****
February	851.72 ± 11.09	1003.29 ± 29.17 **(17.79)**	1492.58 ± 13.67 **(75.24)****	802.41 ± 12.096 **(5.79)**	685.76 ± 27.68 **(19..48)***	515.82 ± 30.56 **(39.43)****

TABLE 35

Sugar Contents (μ gm g^{-1} fr. wt.) of the Leaves of *M.alba* Growing on Different Concentrations of Fly Ash Amended Soil During Different Stages of Growth

	Control	*5%*	*10%*	*20%*	*25%*	*30%*
Stage 1	1875.24	2098.52	2478.35	1745.43	1423.86	1243.70
	±	±	±	±	±	±
	40.67	41.92	52.92	35.55	30.44	26.45
		(11.91)	**(31.16)****	**(6.92)**	**(24.07)***	**(33.68)****
Stage 2	1277.34	1542.45	1786.31	1232.89	935.12	803.45
	±	±	±	±	±	±
	42.24	31.33	41.32	37.35	38.45	21.33
		(20.75)*	**(39.85)****	**(3.48)***	**(26.79)****	**(37.09)****
Stage 3	713.93	857.95	1108.87	672.74	546.90	429.20
	±	±	±	±	±	±
	20.03	36.54	29.82	16.85	33.56	29.26
		(20.17)*	**(55.31)****	**(5.76)**	**(23.39)***	**(39.88)****

TABLE 36

NR Activity (μ mole g^{-1} fr. wt. $hr.^{-1}$) in the Leaves of *C. dichotoma* Growing on Different Concentration of fly ash amended Soil During an Annual Cycle of Three Consecutive Years.

	Control	*5%*	*10%*	*20%*	*25%*	*30%*
March	4.43 ± 0.72	6.65 ± 0.44 **(50.12)****	4.29 ± 0.63 **(3.16)**	3.87 ± 0.36 **(12.64)**	2.84 ± 0.35 **(35.89)****	2.56 ± 0.42 **(42.21)****
April	4.81 ± 0.34	6.43 ± 0.49 **(33.67)****	4.68 ± 0.32 **(2.71)**	3.95 ± 0.23 **(17.87)***	3.08 ± 0.25 **(35.97)****	2.89 ± 0.42 **(39.92)****
May	3.91 ± 0.29	5.53 ± 0.67 **(41.43)****	4.59 ± 0.83 **(17.39)***	3.58 ± 0.42 **(8.43)**	2.63 ± 0.25 **(32.75)****	2.39 ± 0.07 **(38.87)****
June	5.92 ± 0.58	7.11 ± 0.64 **(20.11)***	5.26 ± 0.82 **(11.15)**	4.36 ± 0.75 **(26.35)****	3.76 ± 0.41 **(36.48)****	3.01 ± 0.28 **(49.16)****
July	3.14 ± 0.46	5.46 ± 0.29 **(73.89)***	2.66 ± 0.54 **(15.29)**	2.57 ± 0.21 **(18.15)***	2.37 ± 0.25 **(24.52)**	1.98 ± 0.43 **(36.94)****
August	2.92 ± 0.32	5.11 ± 0.38 **(0.75)****	2.26 ± 0.29 **(22.61)***	2.12 ± 0.05 **(27.39)****	1.71 ± 0.03 **(41.43)****	1.36 ± 0.07 **(53.42)****
September	2.97 ± 0.72	4.76 ± 0.55 **(60.27)****	2.16 ± 0.42 **(27.27)****	1.97 ± 0.07 **(33.68)****	1.69 ± 0.15 **(43.09)****	1.54 ± 0.28 **(48.15)****
October	2.48 ± 0.68	4.15 ± 0.35 **(67.34)****	2.21 ± 0.09 **(10.89)**	1.86 ± 0.08 **(25.11)***	1.78 ± 0.09 **(28.23)****	1.09 ± 0.05 **(56.45)****
November	2.76 ± 0.14	3.74 ± 0.66 **(35.51)****	2.17 ± 0.29 **(21.38)***	1.73 ± 0.05 **(37.23)****	1.68 ± 0.12 **(39.13)****	1.54 ± 0.13 **(44.22)****
December	1.16 ± 0.08	1.63 ± 0.25 **(40.51)****	1.94 ± 0.07 **(18.97)***	1.89 ± 0.12 **(23.28)***	0.91 ± 0.05 **(21.5)***	0.64 ± 0.03 **(44.83)****
January	0.95 ± 0.28	1.55 ± 0.17 **(63.16)****	0.86 ± 0.15 **(9.47)**	0.83 ± 0.04 **(12.63)**	0.64 ± 0.07 **(32.63)****	0.56 ± 0.09 **(41.05)****
February	1.86 ± 0.36	2.98 ± 0.23 **(60.21)****	1.75 ± 0.26 **(5.91)**	1.62 ± 0.13 **(12.93)**	1.45 ± 0.18 **(22.04)***	1.39 ± 0.16 **(25.27)***

TABLE 37

NR Activity (μ mole g^{-1} fr. wt. hr^{-1}) in the Leaves of *C. dichotoma* Growing on Different Concentrations of Fly Ash Amended Soil During Different Stages of Growth

	Control	*5%*	*10%*	*20%*	*25%*	*30%*
Stage 1	4.76 ± 0.48	6.43 ± 0.56 **(34.87)****	4.70 ± 0.65 **(1.31)**	3.94 ± 0.44 **(17.35)***	3.07 ± 0.31 **(35.45)****	2.71 ± 0.31 **(43.10)****
Stage 2	2.87 ± 0.54	4.87 ± 0.39 **(69.24)****	2.32 ± 0.33 **(19.28)***	2.13 ± 0.1 **(25.98)****	1.88 ± 0.13 **(34.40)***	1.49 ± 0.21 **(48.22)***
Stage 3	1.68 ± 0.21	2.47 ± 0.32 **(47.10)****	1.43 ± 0.19 **(15.15.01)**	1.26 ± 0.08 **(24.66)***	1.17 ± 0.11 **(30.46)****	1.03 ± 0.10 **(38.63)****

TABLE 38

NR Activity (μ mole g^{-1} fr. wt. hr^{-1}) in the Leaves of *M. alba* Growing on Different Concentrations of Fly Ash Amended Soil During an Annual Cycle of Three Consecutive Years

	Control	*5%*	*10%*	*20%*	*25%*	*30%*
March	5.95 ± 0.52	7.84 ± 1.17 (13.76)**	8.25 ± 0.64 (38.65)**	5.24 ± 0.78 (11.93)	4.23 ± 0.56 (28.91)**	3.87 ± 0.49 (34.95)**
April	6.47 ± 0.58	8.39 ± 0.71 (29.67)**	9.26 ± 0.23 (43.12)**	5.34 ± 0.48 (17.47)*	4.61 ± 0.39 (28.74)**	3.54 ± 0.65 (45.28)*
May	6.06 ± 0.81	7.96 ± 0.49 (31.35)**	8.37 ± 0.92 (38.11)**	5.88 ± 0.35 (2.97)	4.53 ± 0.38 (25.24)*	3.42 ± 0.27 (43.39)**
June	7.36 ± 0.26	8.68 ± 00.73 (17.93)*	9.84 ± 0.85 (33.69)**	6.57 ± 0.72 (10.73)	5.07 ± 0.43 (31.11)**	4.73 ± 0.27 (35.73)**
July	5.34 ± 0.39	7.47 ± 0.56 (39.88)**	8.39 ± 1.37 (57.11)**	5.25 ± 0.91 (1.68)	3.61 ± 0.43 (32.39)**	3.14 ± 0.57 (41.19)**
August	4.77 ± 0.36	6.67 ± 0.31 (39.83)**	7.02 ± 0.93 (47.16)**	3.69 ± 0.69 (22.64)*	3.21 ± 0.62 (32.71)**	2.62 ± 0.09 (45.02)**
September	4.45 ± 0.96	5.69 ± 0.72 (27.86)**	7.33 ± 0.36 (64.71)	3.24 ± 0.26	3.13 ± 0.43 (27.19)**	2.59 ± 0.27 (29.66)**
(41.97)**						
October	3.43 ± 0.73	4.69 ± 0.62 (36.73)**	5.35 ± 0.68 (55.97)**	2.87 ± 0.51 (16.32)*	1.89 ± 0.47 (44.89)**	1.75 ± 0.52 (48.97)**

TABLE 39

NR Activity (μ mole g^{-1} fr. wt. hr^{-1}) in the Leaves of *M. alba* Growing on Different Concentrations of Fly Ash Amended Soil During Different Stages of Growth

	Control	*5%*	*10%*	*20%*	*25%*	*30%*
STAGE 1	6.46 ± 0.54	8.21 ± 0.77 (27.20)**	8.93 ± 0.66 (38.23)**	5.75 ± 0.58 (10.87)	4.61 ± 0.44 (28.63)**	3.89 ± 0.42 (39.74)**
STAGE 2	4.49 ± 0.61	6.13 ± 0.55 (36.`9)**	7.02 ± 0.83 (56.14)**	3.76 ± 0.59 (16.34)*	2.96 ± 0.48 (34.18)**	2.52 ± 0.36 (32.85)**
STAGE 3	1.82 ± 0.20	2.50 ± 0.35 (36.19)**	3.10 ± 0.57 (69.11)**	1.58 ± 0.14 (13.61)	1.26 ± 0.13 (31.42)**	1.06 ± 0.16 (42.31)**

TABLE 40

Nitrate Content (μ mole g⁻¹ fr. wt.) in the Leaves of *C. dichotoma* Growing on Different Concentrations of Fly Ash Amended Soil During an Annual Cycle of Three Consecutive Years

	Control	*5%*	*10%*	*20%*	*25%*	*30%*
March	1.54 ± 0.13 0.07	1.04 ± 0.12 **(32.46)****	1.39 ± 0.18 **(9.74)**	1.61 ± 0.24 **(4.54)**	2.09 ± 0.17 **(35.71)****	2.82 ± **(83.16)****
April	2.37 ± 0.68	1.97 ± 0.17 **(16.87)***	2.24 ± 0.29 **(5.48)**	2.87 ± 0.18 **(21.09)***	2.92 ± 0.38 **(23.21)***	4.31 ± 0.52 **(81.56)****
May	2.78 ± 0.47	1.63 ± 0.35 **(41.72)****	0.45 ± 0.28 **(11.87)**	2.96 ± 0.31 **(6.47)**	3.36 ± 0.58 **(20.86)***	5.27 ± 0.33 **(89.57)****
June	2.17 ± 0.38	1.83 ± 0.06 **(15.67)***	2.26 ± 0.17 **(4.15)**	2.48 ± 0.32 **(14.29)**	2.63 ± 0.26 **(21.19)***	3.73 ± 0.14 **(71.88)****
July	(3.69) ± 0.42	2.88 ± 0.59 **(21.95)***	3.48 ± 0.36 **(5.69)**	3.85 ± 0.28 **(4.37)**	4.61 ± 0.56 **24.93)***	(5.76) ± 0.37 **(56.09)****
August	3.84 ± 0.35	2.96 ± 0.46 **(22.91)***	3.37 ± 0.19 **(12.24)**	4.04 ± 0.68 **(5.21)**	4.95 ± 0.25 **(28.91)****	6.38 ± 0.51 **(66.15)****
September	2.48 ± 0.19	1.82 ± 0.24 **(26.61)****	2.78 ± 0.56 **(12.09)**	3.38 ± 0.64 **(36.29)****	3.59 ± 0.48 **(44.76)****	4.73 ± 0.39 **(90.73)****
October	4.39 ± 0.216	3.36 ± 0.28 **(23.46)***	3.24 ± 0.43 **(3.42)**	5.28 ± 0.78 **(22.51)***	5.76 ± 0.19 **(31.21)****	7.47 ± 0.53 **(70.16)****
November	3.91 ± 0.53	2.93 ± 0.34 **(25.06)***	4.27 ± 0.46 **(9.21)**	4.53 ± 0.47 **(15.86)***	5.27 ± 0.74 **(34.78)***	6.58 ± 0.18 **(68.28)****
December	4.62 ± 0.41	3.74 ± 0.32 **(19.0%)***	4.31 ± 0.25 **(6.71)**	5.85 ± 0.47 **(26.63)****	6.25 ± 0.61 **(35.28)****	7.95 ± 0.33 **(72.07)****
January	3.12 ± 0.46	2.07 ± 0.08 **(33.65)((**	2.37 ± 0.38 **(24.04)***	3.47 ± 0.89 **(11.22)**	4.28 ± 0.53 **(37.18)****	5.52 ± 0.35 **(76.92)****
February	2.94 ± 0.21	2.06 ± 0.11 **(29.93)****	2.48 ± 0.36 **(15.65)****	3.37 ± 0.42 **(14.63)**	3.64 ± 0.74 **(23.81)****	5.48 ± 0.86 **(86.39)****

TABLE 41

NR content (μ mole g^{-1} fr. wt. hr^{-1}) in the Leaves of *C. dichotoma* Growing on Different Concentrations of Fly Ash Amended Soil During Different Stages of Growth

	Control	*5%*	*10%*	*20%*	*25%*	*30%*
STAGE 1	2.21 ± 0.41	1.61 ± 0.16 **(27.09)****	2.08 ± 0.21 **(5.87)**	2.48 ± 0.24 **(11.96)**	2.75 ± 0.36 **(24.15)***	4.03 ± 0.29 **(82.05)****
STAGE 2	3.61 ± 0.31	2.75 ± 0.39 **(23.47)***	3.46 ± 0.38 **(3.68)**	4.16 ± 0.59 **(15.62)***	4.72 ± 0.37 **(31.31)****	6.08 ± 0.45 **(69.02)****
STAGE 3	3.64 ± 0.40	2.70 ± 0.21 **(25.97)****	3.35 ± 0.36 **(7.95)**	4.30 ± 0.56 **(18.02)***	4.86 ± 0.65 **(33.24)****	6.38 ± 0.43 **(74.98)****

TABLE 42

Nitrate Content (μ mole g^{-1} fr. wt.) in the Leaves of *M. alba* Growing on Different Concentrations of Fly Ash Amended Soil During an Annual Cycle of Three Consecutive Years

	Control	*5%*	*10%*	*20%*	*25%*	*30%*
March	1.34 ± 0.42	1.25 ± 0.45 (6.72)	1.18 ± 0.38 (11.94)	1.78 ± 0.31 (32.84)**	2.76 ± 0.32 105.97)**	3.09 ± 0.14 (130.59)**
April	3.85 ± 0.17	2.94 ± 0.48 (23.64)*	2.79 ± 0.54 (27.53)**	4.37 ± 0.28 (13.51)	5.61 ± 0.79 (45.71)**	6.57 ± 0.85 (70.65)**
May	2.22 ± 0.53	1.76 ± 0.34 (20.72)*	1.58 ± 0.11 (28.83)**	2.64 ± 0.32 (18.91)*	3.78 ± 0.29 (70.27)**	4.07 ± 0.12 (83.35)**
June	3.24 ± 0.78	2.32 ± 0.32 (28.39)**	2.04 ± 0.61 (37.04)**	3.36 ± 0.92 (3.71)	4.58 ± 0.33 (41.35)**	5.97 ± 0.72 (84.25)**
July	5.58 ± 0.12	3.76 ± 0.29 (32.62)**	3.26 ± 0.45 (41.58)**	5.07 ± 0.67 (9.14)	6.38 ± 0.74 (14.34)	7.25 ± 0.56 (29.93)**
August	3.19 ± 0.21	2.48 ± 0.46 (22.26)*	2.44 ± 0.38 (23.51)*	3.72 ± 0.57 (17.24)*	4.95 ± 0.34 (55.17)**	6.03 ± 0.91 (89.02)**
September	1.92 ± 0.83	1.86 ± 0.19 (3.12)	1.42 ± 0.32 (26.04)**	2.37 ± 0.41 (23.43)*	3.56 ± 0.66 (85.42)**	3.64 ± 0.16 (89.58)**
October	5.47 ± 0.42	4.75 ± 0.64 (13.16)	4.51 ± 0.42 (17.55)*	6.23 ± 0.97 (13.89)	6.67 ± 0.3 (21.93)*	7.98 ± 1.13 (45.86)**
November	4.62 ± 0.45	3.86 ± 0.67 (16.45)*	3.34 ± 0.28 (27.71)**	4.78 ± 0.83 (3.46)	5.71 ± 0.52 (23.59)*	6.52 ± 0.36 (41.25)**
December	6.45 ± 0.43	5.05 ± 0.73 (21.75)**	4.15 ± 0.29 (34.68)**	6.95 ± 0.23 (7.75)	7.37 ± 0.66 (14.26)	8.67 ± 0.41 (34.41)**
January	5.24 ± 0.45	4.25 ± 0.13 (18.89)*	4.03 ± 0.21 (23.09)*	4.95 ± 0.81 (5.53)	6.24 ± 0.49 (19.08)*	7.43 ± 0.24 (41.79)**
February	4.35 ± 0.79	3.59 ± 0.41 (17.47)*	2.85 ± 0.16 (34.48)**	4.71 ± 0.42 (8.27)	5.63 ± 0.35 (29.42)**	6.78 ± 0.22 (55.86)**

TABLE 43

Nitrate Content (μ mole g^{-1} fr. wt.) in the Leaves of *M. alba* Growing on Different Concentrations of Fly Ash Amended Soil During Different Stages of Growth

	Control	*5%*	*10%*	*20%*	*25%*	*30%*
STAGE 1	2.56 = 0.47 **(22.34)***	2.06 ± 0.39	1.89 ± 0.41 **(28.73)****	3.03 ± 0.45 **(14.08)**	4.18 ± 0.43 **(57.09)****	4.92 ± 0.45 **(84.98)****
STAGE 2	4.04 ± 0.39	3.31 ± 0.39 **(20.48)***	2.90 ± 0.39 **(28.03)****	4.35 ± 0.65 **(7.74)**	5.39 ± 0.51 **(33.42)****	6.22 ± 0.69 **(54.08)****
STAGE 3	5.16 ± 0.53	4.18 ± 00.48 **(18.93)****	3.59 ± 0.23 **(30.45)****	5.34 ± 0.57 **(3.53)**	6.23 ± 0.50 **(20.76)***	7.35 ± 0.31 **(42.30)****

TABLE 44

Protein Content (mg g^{-1} fr. wt.) of the Leaves of *C. dichotoma* Growing on Different Concentrations of Fly Ash Amended Soil During an Annual Cycle of Three Consecutive Years

	Control	*5%*	*10%*	*20%*	*25%*	*30%*
March	4.26 ± 0.64	6.08 ± 0.48 **(42.72)****	3.45 ± 0.65 **(19.01)***	3.29 ± 0.97 **(22.76)****	2.81 ± 0.45 **(34.03)****	2.49 ± 0.52 **(41.54)****
April	5.51 ± 1.36	7.18 ± 1.05 **(30.31)****	5.59 ± 0.63 **(1.45)**	4.41 ± 0.71 **(19.96)***	3.67 ± 0.95 **(33.39)****	3.25 ± 0.93 **(41.02)****
May	7.19 ± 1.41	9.37 ± 1.77 **(30.32)****	6.83 ± 1.13 **(5.07)**	5.81 ± 0.78 **(19.19)***	5.69 ± 0.82 **(20.86)***	4.56 ± 0.63 **(36.57)****
June	6.09 ± 1.08	8.84 ± 1.11 **(45.16)****	5.84 ± 0.92 **(4.11)**	5.37 ± 0.33 **(11.83)**	3.95 ± 0.78 **(35.14)****	3.75 ± 0.31 **(38.42)****
July	6.57 ± 0.73	9.07 ± 1.51 **(38.05)****	6.92 ± 0.78 **(5.32)**	5.66 ± 0.41 **(13.85)**	4.35 ± 0.66 **(33.79)****	4.28 ± 0.46 **(34.85)****
August	4.51 ± 0.72	6.95 ± 0.87 **(54.12)****	4.37 ± 0.58 **(3.10)**	3.72 ± 0.43 **(17.52)***	2.76 ± 0.31 **(38.81)****	2.64 ± 0.38 **(41.46)****
September	2.71 ± 0.51	3.67 ± 0.96 **(35.42)****	2.02 ± 0.65 **(25.46)***	0.87 ± 0.69 **(30.97)****	1.71 ± 0.15 **(36.93)****	1.56 ± 0.23 **(42.44)****
October	3.26 ± 0.43	5.89 ± 0.52 **(80.68)****	3.48 ± 0.28 **(6.75)**	3.01 ± 0.71 **(7.67)**	2.52 ± 0.46 **(22.69)***	1.97 ± 0.37 **(39.57)****
November	1.29 ± 0.21	1.71 ± 0.18 **(32.56)****	0.99 ± 0.06 **(23.26)***	0.87 ± 0.04 **(32.56)****	0.82 ± 0.17 **(36.43)****	0.75 ± 0.16 **(41.86)****
December	0.83 ± 0.16	1.25 ± 0.21 **(50.62)****	0.89 ± 0.18 **(7.23)**	0.65 ± 0.03 **(21.68)***	0.53 ± 0.08 **(36.14)****	0.48 ± 0.05 **(42.17)****
January	0.62 ± 0.06	0.95 ± 0.05 **(53.23)****	0.64 ± 0.13 **(3.23)**	0.58 ± 0.07 **(6.45)**	0.47 ± 0.05 **24.19)***	0.42 ± 0.07 **(32.26)****
February	0.25 ± 0.81	3.96 ± 0.34 **(76.17)****	1.96 ± 0.25 **(12.89)**	1.77 ± 0.23 **(21.33)***	1.38 ± 0.07 **(38.67)****	1.29 ± 0.15 **(42.67)****

TABLE 45

Protein Content (mg g^{-1} fr. wt.) of the Leaves of *C. dichotoma* Growing on Different Concentrations of Fly Ash Amended Soil During Different Stages of Growth

	Control	*5%*	*10%*	*20%*	*25%*	*30%*
STAGE 1	5.76 ± 1.12	7.86 ± 1.35 **(36.53)****	5.42 ± 0.83 **(5.81)**	4.72 ± 0.69 **(18.09)***	4.03 ± 0.75 **(30.06)****	3.51 ± 0.59 **(39.04)****
STAGE 2	4.26 ± 0.59	6.39 ± 0.96 **(50.02)****	4.19 ± 0.57 **(1.52)**	3.56 ± 0.56 **(16.36)***	2.83 ± 0.39 **(33.49)****	2.61 ± 0.36 **(38.71)****
STAGE 3	1.24 ± 0.31	1.96 ± 0.19 **(57.71)****	1.12 ± 0.15 **(10.22)**	0.96 ± 0.09 **(22.44)***	0.8 ± 0.09 **(35.87)****	0.73 ± 0.11 **(41.08)****

TABLE 46

Protein Content (mg g^{-1} fr. wt.) of the Leaves of *M. alba* Growing on Different Concentrations of Fly Ash Amended Soil During an Annual Cycle of Three Consecutive Years

	Control	*5%*	*10%*	*20%*	*25%*	*30%*
March	4.12 ± 0.56	5.57 ± 0.89 **(35.19)****	6.02 ± 1.03 **(46.16)****	3.666 ± 0.67 **(11.62)**	.323 ± 0.49 **(22.58)***	2.78 ± 0.47 **(32.52)***
April	5.94 ± 1.05	7.14 ± 0.72 **(20.28)***	8.84 ± 1.95 **(48.82)****	5.46 ± 0.86 **(8.09)**	4.22 ± 0.84 **(28.96)****	3.29 ± 0.58 **(44.61)****
May	3.65 ± 0.19	4.11 ± 0.32 **(12.67)**	5.49 ± 0.27 **(50.51)****	3.03 ± 0.46 **(16.98)***	2.91 ± 0.44 **(20.39)***	1.93 ± 0.09 **(47.12)****
June	3.27 ± 0.45	4.07 ± 0.67 **(24.46)***	5.58 ± 0.56 **(70.64)****	2.85 ± 0.64 **(12.84)**	2.64 ± 0.32 **(19.27)***	1.94 ± 0.23 **(40.67)****
July	2.27 ± 0.16	3.06 ± 0.42 **(34.81)****	3.59 ± 0.14 **(58.14)****	2.05 ± 0.09 **(9.69)**	1.51 ± 0.04	1.45 ± 0.08 **(36.12)****
August	2.05 ± 0.09	2.83 ± 0.18 **(38.04)****	3.77 ± 0.25 **(83.91)****	1.64 ± 0.23 **(20.63)***	1.12 ± 0.17 **(40.33)****	1.16 ± 0.07 **(43.42)****
September	2.69 ± 0.73	3.43 ± 0.42 **(27.51)****	4.07 ± 1.02 **(51.31)****	2.48 ± 0.57 **(7.86)**	2.24 ± 0.56 **(16.72)***	1.65 ± 0.22 **(38.66)****
October	1.97 ± 0.15	2.45 ± 0.32 **(24.36)***	3.23 ± 0.28 **(63.95)****	1.62 ± 0.16 **(17.76)***	1.35 ± 0.09 **(31.47)****	1.07 ± 0.07 **(45.66)****
November	1.54 ± 0.25	2.33 ± 0.36 **(51.29)****	2.72 ± 0.43 **(76.62)****	1.17 ± 0.08 **(24.02)***	0.95 ± 0.03 **(38.32)****	0.92 ± 0.08 **(40.25)****
December	0.69 ± 0.05	0.94 ± 0.14 **(36.23)****	1.04 ± 0.16 **(50.72)****	0.75 ± 0.06 **(8.69)**	0.52 ± 0.02 **(24.63)***	0.41 ± 0.03 **(40.58)****
January	0.68 ± 0.08	0.82 ± 0.04 **(20.58)***	1.14 ± 0.06 **(67.64)****	0.52 ± 0.07 **(23.52)***	0.43 ± 0.04 **(36.76)****	0.37 ± 0.03 **(45.28)****
February	0.85 ± 0.16	1.08 ± 0.15 **(27.06)****	1.46 ± 0.12 **(71.47)****	0.72 ± 0.03 **(15.29)**	0.64 ± 0.04 **(24.58)***	0.53 ± 0.04 **(36.47)****

TABLE 47

Protein Content (mg g[-1] fr. wt.) of the Leaves of *M. alba* Growing on Different Concentrations of Fly Ash Amended Soil During Sifferent Stages of Growth

	Control	*5%*	*10%*	*20%*	*25%*	*30%*
STAGE 1	4.24 ± 0.56	5.22 ± 0.65 **(23.02)***	6.48 ± 0.95 **(52.71)****	3.72 ± 0.66 **(11.66)**	3.25 ± 0.52 **(23.44)***	2.48 ± 0.34 **(41.46)****
STAGE 2	2.24 ± 0.28	2.24 ± 0.33 **(31.07)****	3.66 ± 0.42 **(63.25)****	1.94 ± 0.26 **(13.25)**	1.58 ± 0.21 **(29.51)****	1.33 ± 0.11 **(40.65)****
STAGE 3	0.94 ± 0.13	1.29 ± 0.17 **(37.5)****	1.59 ± 0.19 **(69.15)****	0.79 ± 0.06 **(15.96)***	0.63 ± 0.03 **(32.45)****	0.56 ± 0.04 **(40.42)****

TABLE 48

Leaf pH of *C. dichotoma* Growing on Different Concentrations of Fly Ash Amended Soil During an Annual Cycle of Three Consecutive Years

	Control	*5%*	*10%*	*20%*	*25%*	*30%*
March	6.79 ± 0.026	6.85 ± 0.04 **(0.88)**	6.87 ± 0.05 **(1.17)**	6.92 ± 0.03 **(1.91)**	6.98 ± 0.03 **(2.79)**	7.09 ± 0.03 **(4.41)**
April	6.92 ± 0.23	6.98 ± 0.14 **(0.86)**	7.04 ± 0.03 **(1.73)**	7.09 ± 0.12 **(2.46)**	7.15 ± 0.07 **(3.32)**	7.12 ± 0.06 **(2.89)**
May	6.83 ± 0.06	6.91 ± 0.07 **(1.17)**	6.99 ± 0.17 **(2.34)**	7.02 ± 0.24 **(2.78)**	7.09 ± 0.06 **3.81**	7.11 ± 0.19 **(4.09)**
June	7.02 ± 0.12	7.08 ± 0.05 **(0.87)**	7.17 ± 0.03 **(2.14)**	7.19 ± 0.03 **(2.42)**	7.21 ± 0.17 **(2.71)**	7.27 ± 0.04 **(3.56)**
July	7.06 ± 0.07	7.12 ± 0.13 **(0.84)**	7.18 ± 0.08 **(1.69)**	7.23 ± 0.07 **2.41)**	7.26 ± 0.09 **(2.83)**	7.29 ± 0.12 **(3.25)**
August	7.01 ± 0.09	7.04 ± 0.03 **(0.43)**	7.09 ± 0.09 **(1.14)**	7.11 ± 0.06 **(1.46)**	7.19 ± 0.13 **(2.56)**	7.25 ± 0.19 **(3.42)**
September	6.96 ± 0.32	7.08 ± 0.02 **(1.72)**	7.09 ± 0.04 **(1.86)**	7.15 ± 0.16 **(2.79)**	7.26 ± 0.07 **(4.31)**	7.24 ± 0.03 **(4.02)**
October	7.02 ± 0.03	7.08 ± 0.04 **(0.85)**	7.13 ± 0.06 **(1.56)**	7.15 ± 0.09 **(1.85)**	7.29 ± 0.04 **(3.84)**	7.32 ± 0.03 **(4.27)**
November	7.09 ± 0.12	7.14 ± 0.07 **(0.71)**	7.15 ± 0.19 **(0.85)**	7.21 ± 0.08 **(1.69)**	7.21 ± 0.03 **(1.69)**	7.26 ± 0.05 **(2.39)**
December	7.11 ± 0.03	7.12 ± 0.04 **(0.14)**	7.12 ± 0.09 **(0.14)**	7.19 ± 0.04 **(1.12)**	7.23 ± 0.07 **(1.68)**	7.28 ± 0.06 **(2.39)**
January	7.11 ± 0.05	7.16 ± 0.04 **(0.71)**	7.19 ± 0.09 **(1.13)**	7.24 ± 0.07 **(1.82)**	7.25 ± 0.11 **(1.96)**	7.32 ± 0.09 **(2.95)**
February	7.12 ± 0.09	7.19 ± 0.03 **(0.98)**	7.23 ± 0.15 **(1.54)**	7.26 ± 0.07 **(1.96)**	7.27 ± 0.10 **(2.10)**	7.31 ± 0.04 **(2.67)**

TABLE 49

Leaf pH of *C. dichotoma* Growing on Different Concentration of Fly Ash Amended Soil During Different Stages of Growth

	Control	*5%*	*10%*	*20%*	*25%*	*30%*
Stage 1	6.89 ± 0.11	6.95 ± 0.07 (0.94)	7.01 ± 0.07 (1.85)	7.05 ± 0.10 (2.39)	7.10 ± 0.08 (3.16)	7.14 ± 0.08 (3748)
Stage 2	7.01 ± 0.12	7.08 ± 0.05 (0.96)	7.12 ± 0.06 (1.57)	7.16 ± 0.09 (2.10)	7.25 ± 0.08 (3.39)	7.27 ± 0.09 (3.74)
Stage 3	7.10 ± 0.07	7.15 ± 0.04 (0.63)	7.17 ± 0.13 (0.91)	7.22 ± 0.06 (1.65)	7.24 ± 0.07 (1.86)	7.29 ± 0.06 (2.60)

TABLE 50

Leaf pH of *M. alba* Growing on Different Concentrations of Fly Ash Amended Soil during an Annual Cycle of Three Consecutive Years

	Control	*5%*	*10%*	*20%*	*25%*	*30%*
March	6.32 ± 0.05	6.38 ± 0.27 **(0.94)**	6.41 ± 0.06 **(1.42)**	6.45 ± 0.18 **(2.06)**	6.51 ± 0.24 **(3.01)**	7.57 ± 0.06 **(3.95)**
April	6.36 ± 0.02	6.42 ± 0.06 **(0.93)**	6.45 ± 0.03 **(1.41)**	6.52 ± 0.05 **(2.51)**	6.57 ± 0.23 **(3.31)**	6.67 ± 0.18 **(5.03)**
May	6.69 ± 0.36	6.84 ± 0.06 **(2.24)**	6.98 ± 0.05 **(4.34)**	7.98 ± 0.05 **(5.82)**	7.12 ± 0.19 **(6.43)**	7.16 ± 0.05 **(7.02)**
June	6.42 ± 0.06	6.54 ± 0.13 **(1.86)**	6.61 ± 0.34 **(2.95)**	6.68 ± 0.04 **(4.05)**	6.57 ± 0.03 **(2.37)**	6.73 ± 0.05 **(4.82)**
July	6.53 ± 0.04	6.54 ± 0.06 **(0.15)**	6.63 ± 0.15 **1.53)**	6.76 ± 0.44 **(3.52)**	6.78 ± 0.06 **(3.82)**	6.82 ± 0.09 **(4.44)**
August	6.64 ± 0.06	6.74 ± 0.07 **(1.51)**	6.79 ± 0.07 **(2.25)**	6.82 ± 0.06 **(2.71)**	6.87 ± 0.14 **(3.46)**	6.85 ± 0.25 **(3.16)**
September	7.05 ± 0.11	7.12 ± 0.27 **(0.99)**	7.13 ± 0.09 **(1.14)**	7.24 ± 0.22 **(2.69)**	7.21 ± 0.06 **(2.27)**	7.29 ± 0.15 **(3.41)**
October	7.11 ± 0.09	7.12 ± 0.06 **(0.1)**	7.16 ± 0.07 **(0.73)**	7.20 ± 0.16 **(1.260**	7.24 ± 0.15 **(1.82)**	7.27 ± 0.06 **(2.25)**
November	7.17 ± 0.05	7.20 ± 0.12 **(0.42)**	7.24 ± 0.07 **(0.97)**	7.31 ± 0.25 **(1.95)**	7.37 ± 0.09 **(2.78)**	7.38 ± 0.14 **(2.92)**
December	6.78 ± 0.07	6.77 ± 0.14 **(0.15)**	6.82 ± 0.27 **(0.58)**	6.87 ± 0.29 **(1.32)**	6.92 ± 0.05 **(2.06)**	6.93 ± 0.09 **(2.21)**
January	7.23 ± 0.25	7.28 ± 0.06 **(0.69)**	7.34 ± 0.07 **(1.52)**	7.36 ± 0.08 **(1,79)**	7.40 ± 0.16 **(2.35)**	7.42 ± 0.31 **(2.62)**
February	7.34 ± 0.17	7.41 ± 0.06 **(0.95)**	7.41 ± 0.14 **(0.95)**	7.46 ± 0.08 **(1.64)**	7.52 ± 0.32 **(2.45)**	7.56 ± 0.03 **(2.99)**

TABLE 51

Leaf pH of *M. alba* Growing on Different Concentrations of Fly Ash Amended Soil during Different Stages of Growth

	Control	5%	10%	20%	25%	30%
STAGE 1	6.44 ± 0.12	6.54 ± 0.13 **(1.51)**	6.61 ± 0.12 **(2.56)**	6.68 ± 0.17 **(3.64)**	6.69 ± 0.17 **(3.79)**	6.78 ± 0.08 **(5.23)**
STAGE 2	6.83 ± 0.07	6.8 ± 0.11 **(0.69)**	6.92 ± 0.09 **(1.39)**	7.01 ± 0.17 **(2.52)**	7.02 ± 0.10 **(2.81)**	7.05 ± 0.13 **(3.29)**
STAGE 3	7.13 ± 0.13	7.16 ± 0.09 **(0.49)**	7.20 ± 0.13 **(1.01)**	7.25 ± 0.12 **(1.68)**	7.30 ± 0.15 **(2.41)**	7.32 ± 0.14 **(2.69)**

TABLE 52

Nitrogen Content (mg g⁻¹ dry wt.) of the Leaves of *C. dichotoma* Growing on Different Concentrations of Fly Ash Amended Soil during an Annual Cycle of Three Consecutive Years

	Control	*5%*	*10%*	*20%*	*25%*	*30%*
March	2.27 ± 0.42	2.34 ± 0.31 **(3.08)**	2.21 ± 0.32 **(2.64)**	2.09 ± 0.37 **(7.93)**	1.57 ± 0.45 **(30.83)****	1.45 ± 0.09 **(36.12)****
April	3.91 ± 0.08	4.03 ± 0.74 **(3.06)**	3.98 ± 0.59 **(1.79)**	3.75 ± 0.12 **94.09)**	3.22 ± 0.71 **(15.08)**	3.06 ± 0.31 **(21.73)***
May	2.71 ± 0.54	2.84 ± 0.32 **(4.79)**	2.44 ± 0.25 **(9.96)**	1.59 ± 0.16 **(41.32)****	1.64 ± 0.46 **(39.48)****	1.51 ± 0.27 **(44.28)****
June	1.84 ± 0.09	2.01 ± 0.19 **(9.23)**	1.62 ± 0.05 **(11.95)**	1.48 ± 0.08 **(19.56)***	1.23 ± 0.15 **(33.15)****	1.06 ± 0.03 **(42.39)****
July	3.18 ± 0.29	3.29 ± 0.47 **(3.46)**	2.88 ± 0.05 **(9.43)**	2.34 ± 0.06 **(26.41)***	2.39 ± 0.15 **(24.82)***	1.92 ± 0.14 **(39.64)****
August	2.82 ± 0.09	3.18 ± 1.02 **(12.76)**	3.12 ± 0.48 **(10.64)**	2.62 ± 0.57 **(7.09)**	2.14 ± 0.06 **(24.11)***	1.83 ± 0.42 **(35.10)****
September	4.46 ± 1.23	4.71 ± 0.64 **(5.65)**	4.11 ± 0.82 **(7.85)**	3.75 ± 0.45 **(15.91)***	3.51 ± 0.21 **(21.34)***	3.11 ± 0.14 **(30.26)****
October	4.31 ± 0.46	4.48 ± 0.36 **(3.94)**	4.26 ± 0.67 **(1.16)**	4.06 ± 1.05 **(5.84)**	3.95 ± 0.37 **(8.35)**	3.39 ± 0.08 **(21.34)***
November	4.59 ± 1.08	4.82 ± 0.13 **(5.01)**	4.12 ± 0.52 **(10.23)**	3.45 ± 0.63 **(24.86)***	3.19 ± 0.07 **(30.59)****	2.62 ± 0.08 **(42.91)****
December	2.52 ± 0.13	2.72 ± 0.36 **(7.93)**	2.39 ± 0.05 **(5.15)**	1.79 ± 0.14 **(28.96)****	2.04 ± 0.17 **(19.04)***	1.51 ± 0.12 **(40.07)****
January	2.51 ± 0.23	2.82 ± 0.26 **(12.35)**	2.46 ± 0.37 **(1.99**	2.21 ± 0.06 **(11.95)**	1.96 ± 0.23 **(21.91)***	1.76 ± 0.04 **(29.68)****
February	1.81 ± 0.48	2.16 ± 0.07 **(19.33)**	1.87 ± 0.25 **(3.31)**	1.56 ± 0.05 **(13.81)**	1.48 ± 0.12 **(18.23)***	1.41 ± 0.25 **(22.09)***

TABLE 53

Nitrogen Content (mg g^{-1} dry wt.) of the Leaves of *C. dichotoma* Growing on Different Concentrations of Fly Ash Amended Soil During Different Stages of Growth

	Control	*5%*	*10%*	*20%*	*25%*	*30%*
STAGE 1	2.63 ± 0.23	2.80 ± 0.39 **(4.57)**	2.56 ± 0.30 **(4.47)**	2.22 ± 0.18 **(16.96)***	1.94 ± 0.44 **(27.68)****	1.77 ± 0.17 **(34.02)**
STAGE 2	3.69 ± 0.51	3.91 ± 0.62 **(6.03)**	3.59 ± 0.50 **(2.71)**	3.19 ± 0.53 **(13.54)**	2.99 ± 0.19 **(18.82)***	2.56 ± 0.19 **(30.60)****
STAGE 3	2.85 ± 0.48	3.13 ± 0.20 **(9.54)**	2.71 ± 0.29 **(5.16)**	2.25 ± 0.22 **(21.17)***	2.16 ± 0.14 **(24.15)***	1.82 ± 0.12 **(36.13)****

TABLE 54

Nitrogen Content (mg g[-1] dry wt.) of the Leaves of *M. alba* Growing on Different Concentrations of Fly Ash Amended Soil during an Annual Cycle of Three Consecutive Years

	Control	*5%*	*10%*	*20%*	*25%*	*30%*
March	2.15 ± 0.36	2.39 ± 0.34 **(11.16)**	2.43 ± 0.41 **(13.02)**	1.81 ± 0.23 **(15.81)***	1.73 ± 0.27 **(19.53)***	1.56 ± 0.42 **(27.44)****
April	3.01 ± 1.11	3.09 ± 0.82 **(2.65)**	3.18 ± 0.69 **(5.65)**	2.97 ± 0.56 **(1.32)**	2.53 ± 0.39 **(15.94)***	2.57 ± 0.43 **(14.61)**
May	2.46 ± 0.29	2.55 ± 0.31 **(3.65)**	2.59 ± 0.24 **(5.28)**	2.09 ± 0.16 **(15.04)**	1.99 ± 0.49 **(19.01)***	1.82 ± 0.37 **(26.01)****
June	4.45 ± 0.24	4.64 ± 0.73 **(4.27)**	4.75 ± 0.87 **(6.74)**	3.69 ± 0.43 **(17.07)***	2.79 ± 0.26 **(37.31)****	2.41 ± 0.17 **(45.84)****
July	3.64 ± 0.42	3.62 ± 0.23 **(0.39)**	4.14 ± 1.52 **(13.73)**	3.39 ± 0.95 **(6,.86)**	2.77 ± 0.65 **(23.91)***	2.31 ± 0.52 **(36.53)****
August	5.42 ± 0.74	5.62 ± 1.96 **(3.69)**	5.84 ± 0.41 **(7.74)**	4.15 ± 0.35 **823.43)***	4.27 ± 0.53 **(21.27)***	3.98 ± 0.24 **(26.56)****
September	3.47 ± 1.02	3.85 ± 0.49 **(10.95)**	3.95 ± 0.13 **(13.83)**	2.77 ± 0.64 **(20.17)***	2.53 ± 0.42 **(27.08)****	2.28 ± 0.07 **(34.29)****
October	4.74 ± 0.95	4.93 ± 0.32 **(4.01)**	5.06 ± 1.11 **(6.75)**	3.54 ± 0.43 **(25.31)****	3.28 ± 0.54 **(30.81)****	2.69 ± 0.35 **(43.24)****
November	4.25 ± 0.42	4.23 ± 0.57 **(0.47)**	4.78 ± 0.83 **(12.47)**	4.16 ± 0.63 **(2.12)**	3.98 ± 0.77 **(6.35)**	3.25 ± 0.98 **(23.52)***
December	2.51 ± 0.12	2.76 ± 0.34 **(9.96)**	2.93 ± 0.46 **(16.73)***	2.11 ± 0.39 **(15.93)***	1.84 ± 0.45 **(26.69)****	1.75 ± 0.09 **(30.27)****
January	3.03 ± 0.54	3.14 ± 0.36 **(3.63)**	3.46 ± 0.39 **(14.19)**	2.74 ± 0.62 **(9.57)**	2.21 ± 0.09 **(27.06)***	0.02 ± 0.29 **(33.34)****
February	3.65 ± 0.57	3.85 ± 0.28 **(5.47)**	3.94 ± 0.46 **(7.94)**	3.31 ± 0.38 **(7.12)**	2.59 ± 0.18 **(29.04)****	2.42 ± 0.57 **(33.69)****

TABLE 55

Nitrogen Content (mg g⁻¹ dry wt.) of the Leaves of *M. alba* Growing on Different Concentrations of Fly Ash Amended Soil during Different Stages of Growth

	Control	*5%*	*10%*	*20%*	*25%*	*30%*
STAGE 1	3.01 ± 0.54	3.16 ± 0.55 **(4.97)**	3.23 ± 0.55 **(7.29)**	2.64 ± 0.34 **(12.51)**	2.26 ± 0.35 **(25.10)***	2.09 ± 0.34 **(30.73)****
STAGE 2	4.31 ± 0.78	4.50 ± 0.75 **(4.34)**	4.74 ± 0.79 **(9.96)**	3.46 ± 0.59 **(19.80)***	3.21 ± 0.53 **(25.59)****	2.81 ± 0.29 **(34.80)****
STAGE 3	3.36 ± 0.41	3.49 ± 0.38 **(4.01)**	3.77 ± 0.53 **(12.42)**	3.15 ± 0.50 **(7.74)**	2.65 ± 0.37 **(20.98)***	2.36 ± 0.48 **(29.76)****

TABLE 56

Sulphur Content (μ gm g^{-1} dry wt.) of the Leaves of *C. dichotoma* Growing on Different Concentrations of Fly Ash Amended Soil during an Annual cycle of Three Consecutive Years

	Control	*5%*	*10%*	*20%*	*25%*	*30%*
March	225.84 ± 36.82	224.65 ± 21.49 **(0.52)**	219.04 ± 17.16 **(3.01)**	214.45 ± 21.49 **(5.04)**	211.29 ± 14.79 **(6.44)**	202.48 ± 15.21 **(10.34)**
April	213.95 ± 27.83	207.92 ± 34.36 **(2.82)**	224.62 ± 19.96 **(4.98)**	218.47 ± 11.54 **(2.11)**	213.31 ± 25.97 **(0.29)**	215.35 ± 12.45 **(0.62)**
May	215.62 ± 24.59	201.64 ± 28.95 **(2.77**	189.85 ± 15.33 **(11.95)**	199.82 ± 25.07 **(7.32)**	170.51 ± 15.07 **(20.92)***	171.67 ± 16.14 **(20.38)***
June	237.32 ± 18.71	210.99 ± 19.28 **(11.09)**	205.23 ± 25.81 **(13.52)**	179.07 ± 19.26 **(24.54)***	183.98 ± 25.43 **(22.47)***	168.27 ± 10.24 **(29.01)****
July	234.33 ± 19.63	250.12 ± 32.91 **(6.73)**	236.54 ± 24.41 **(0.94)**	220.83 ± 17.89 **(5.76)**	239.62 ± 14.38 **(2.27)**	206.95 ± 16.16 **(11.68)**
August	244.67 ± 28.34	236.98 ± 20.46 **(3.14)**	214.17 ± 25.76 **(12.46)**	225.93 ± 19.29 **(7.64)**	221.66 ± 29.17 **(9.41)**	210.26 ± 20.37 **(14.06)**
September	228.34 ± 18.09	197.31 ± 19.14 **(13.59)**	207.22 ± 24.28 **(9.25)**	211.19 ± 17.65 **(7.51)**	203.13 ± 17.34 **(11.04)**	185.41 ± 20.23 **(18.89)***
October	251.72 ± 19.49	235.97 ± 12.64 **(6.25)**	221.58 ± 21.49 **(11.79)**	185.47 ± 20.79 **(26.31)***	217.62 ± 25.45 **(13.54)**	194.65 ± 28.23 **(22.67)***
November	209.12 ± 26.65	211.31 ± 16.45 **(1.05)**	203.34 ± 19.92 **(2.76)**	217.18 ± 28.31 **(3.85)**	184.13 ± 16.15 **(11.95)**	177.32 ± 29.14 **(15.24)**
December	216.75 ± 19.66	204.93 ± 23.37 **(5.45)**	210.96 ± 18.39 **(2.67)**	202.73 ± 16.82 **(6.47)**	190.48 ± 24.68 **(12.11)**	199.64 ± 17.31 **(7.89)**
January	223.54 ± 11.42	214.76 ± 25.67 **(3.93)**	233.93 ± 13.16 **(4.64)**	236.47 ± 22.14 **(5.76)**	205.63 ± 23.22 **(8.01)**	193.26 ± 20.95 **(13.54)**
February	219.64 ± 23.14	227.38 ± 17.52 **(3.52)**	221.17 ± 22.35 **(0.69**	210.98 ± 21.27 **(3.94)**	178.83 ± 19.72 **(18.59)***	168.24 ± 29.28 **(23.41)***

TABLE 57

Sulphur Content (μ gm g[-1] dry wt.) of the Leaves of *C. dichotoma* Growing on Different Concentrations of Fly Ash Amended Soil during Different Stages of Growth

	Control	*5%*	*10%*	*20%*	*25%*	*30%*
STAGE 1	223.18 ± 26.98	213.3 ± 26.02 **(4.43)**	209.68 ± 19.36 **(6.05)**	202.95 ± 19.34 **(9.06)**	194.77 ± 20.31 **(12.73)**	189.46 ± 13.51 **(15.11)**
STAGE 2	239.76 ± 21.38	239.09 ± 21.28 **(4.03)**	219.87 ± 23.98 **(8.29)**	210.86 ± 18.90 **(12.05)**	220.50 ± 21.58 **(8.03)**	199.31 ± 21.24 **(16.87)***
STAGE 3	217.26 ± 20.21	214.59 ± 20.75 **(1.23)**	217.35 ± 18.45 **(0.04)**	216.84 ± 22.13 **(0.09)**	189.76 ± 20.94 **(12.66)**	184.59 ± 24.17 **(15.04)**

TABLE 58

Sulphur Content (μ gm g⁻¹ dry wt.) of the Leaves of *M. alba* Growing on Different Concentrations of Fly Ash Amended Soil during an Annual Cycle of Three Consecutive Years

	Control	*5%*	*10%*	*20%*	*25%*	*30%*
March	216.11 ± 11.28	199.71 ± 15.16 **(7.58)**	186.16 ± 23.14 **(13.85)**	190.19 ± 20.21 **(11.99)**	175.21 ± 20.47 **(18.92)***	173.58 ± 12.45 **(19.67)***
April	242.47 ± 14.36	213.09 ± 27.41 **(4.69)**	257.87 ± 12.32 **(6.35)**	246.14 ± 18.72 **(1.51)**	234.22 ± 20.73 **(3.42)**	230.75 ± 9.17 **(4.83)**
May	233.46 ± 27.09	242.86 ± 16.41 **(4.02)**	250.92 ± 38.07 **(7.47)**	238.77 ± 16.32 **(2.27)**	247.18 ± 14.37 **(5.87)**	241.31 ± 25.94 **(3.36)**
June	237.26 ± 36.23	229.71 ± 26.67 **(3.15)**	224.65 ± 27.09 **(5.31)**	230.31 ± 24.25 **(2.920**	242.05 ± 15.37 **(2.01)**	246.68 ± 35.64 **(3.97)**
July	246.58 ± 34.75	256.88 ± 16.57 **(4.18)**	258.93 ± 37.32 **(5.01)**	261.75 ± 25.48 **(6.15)**	256.78 ± 23.66 **(4.1)**	265.97 ± 15.3 **(7.86)**
August	265. ± 39.16	249.84 ± 10.15 **(5.83)**	261.53 ± 29.54 **(1.43)**	253.41 ± 24.48 **(4.13)**	247.95 ± 27.29 **(6.55)**	258.78 ± 15.43 **(2.47)**
September	258.94 ± 17.54	260.41 ± 28.12 **(0.57)**	271.72 ± 11.08 **(4.93)**	265.71 ± 52.02 **(2.61)**	274.94 ± 41.29 **(6.17)**	261.27 ± 20.44 **(3.98)**
October	240.88 ± 18.12	239.13 ± 27.16 **(0.72)**	248.95 ± 38.11 **(3.35)**	251.62 ± 24.53 **(4.446)**	245.29 ± 17.99 **(1.83)**	234.17 ± 29.43 **(2.78)**
November	189.68 ± 27.37	201.61 ± 30.15 **(6.29)**	211.71 ± 28.78 **(11.61)**	217.04 ± 11.21 **(14.42)**	209.31 ± 18.01 **(10.34)**	214.26 ± 16.96 **(12.95)**
December	171.21 ± 28.18	207.35 ± 19.86 **(21.18)***	198.64 ± 25.73 **(16.02)**	210.09 ± 21.64 **(22.79)**	185.22 ± 20.35 **(8.18)**	189.64 ± 26.27 **(10.76)**
January	196.37 ± 16.64	189.42 ± 20.48 **(3.53)**	184.63 ± 18.55 **(5.97)**	190.18 ± 12.32 **(3.15)**	196.54 ± 17.94 **(0.08)**	207.89 ± 29.47 **(5.87)**
February	209.51 ± 21.16	212.77 ± 27.16 **(1.56)**	197.19 ± 18.78 **(5.88)**	195.86 ± 29.54 **(6.51)**	204.02 ± 21.46 **(2.62)**	121.25 ± 27.31 **(1.31)**

TABLE 59

Sulphur Content (μ gm g⁻¹ dry wt.) of the Leaves of *M. alba* Growing on Different Concentrations of Fly Ash Amended Soil during Different Stages of Growth

	Control	5%	10%	20%	25%	30%
STAGE 1	232.32 ± 22.24	225.86 ± 21.41 **(2.78)**	229.9 ± 25.51 **(1.04)**	226.35 ± 19.87 **(2.57)**	224.66 ± 17.86 **(3.29)**	223.08 ± 20.8 **(3.98)**
STAGE 2	252.93 ± 27.39	251.56 ± 20.50 **(0.54)**	260.28 ± 29.01 **(2.90)**	258.37 ± 31.57 **(2.15)**	256.24 ± 27.55 **(1.31)**	257.04 ± 20.15 **(1.62)**
STAGE 3	191.69 ± 23.33	292.78 ± 24.41 **(5.78)**	198.04 ± 22.96 **(3.31)**	293.29 ± 18.67 **(6.05)**	198.77 ± 19.44 **(3.69)**	206.01 ± 25.00 **(7.46)**

TABLE 60

Photosynthetic Rate (μ mole CO_2 m^{-2} s^{-1}) of the Leaves of *C. dichotoma* Growing on Different Concentrations of Fly Ash Amended Soil during an Annual Cycle of Three Consecutive Years

	Control	*5%*	*10%*	*20%*	*25%*	*30%*
March	17.61 ± 1.86	24.38 ± 2.89 **(29.95)****	13.25 ± 2.06 **19.28)***	10.25 ± 1.931 **(32.56)****	8.32 ± 0.59 **(41.08)****	7.01 ± 0.43 **(46.88)****
April	18.41 ± 2.12	26.42 ± 3.17 **(39.25)****	14.79 ± 2.35 **(17.74)***	13.36 ± 1.87 **(24.74)***	11.72 ± 0.75 **(32.77)****	10.15 ± 1.21 **(40.47)****
May	15.39 ± 0.84	23.35 ± 1.89 **(46.03)****	13.64 ± 1.16 **(9.52)**	9.11 ± 1.09 **(34.15)****	8.94 ± 1.22 **(35.07)****	8.25 ± 0.95 **(38.83)****
June	15.26 ± 0.92	23.32 ± 1.10 **(44.14)****	13.01 ± 0.92 **(12.32)**	8.63 ± 1.74 **(37.02)****	8.49 ± 0.33 **(37.13)****	8.22 ± 0.74 **(38.55)****
July	14.87 ± 1.11	23.53 ± 1.44 **(51.33)****	13.68 ± 1.98 **(7.06)**	9.09 ± 0.03 **(34.26)****	7.72 ± 0.99 **(42.38)****	6.74 ± 0.49 **(48.01)****
August	14.08 ± 1.29	22.73 ± 1.02 **(57.36)****	14.71 ± 1.11 **(4.18)**	93.66 ± 0.93 **(29.31)****	8.17 ± 0.18 **(39.19)****	7.43 ± 0.92 **(44.09)****
September	13.88 ± 1.41	29.57 ± 1.23 **(48.19)****	14.09 ± 1.52 **(1.51)**	8.15 ± 0.94 **(41.28)****	7.24 ± 0.62 **(47.84)****	0.34 ± 0.43 **(52.16)****
October	12.35 ± 0.63	19.68 ± 1.49 **(59.35)****	12.13 ± 1.47 **(1.78)**	7.85 ± 0.96 **(36.44)****	6.98 ± 0.48 **(43.48)****	6.50 ± 0.46 **(47.37)****
November	11.21 ± 1.83	15.62 ± 0.71 **(39.34)****	10.11 ± 0.91 **(9.81)**	7.73 ± 0.89 **(31.04)****	7.45 ± 0.45 **(33.54)****	6.03 ± 0.87 **(46.21)****
December	8.92 ± 0.37	12.41 ± 1.32 **(39.13)****	8.63 ± 0.53 **(3.25)**	5.41 ± 0.76 **(39.24)****	4.23 ± 0.95 **(52.57)****	4.03 ± 0.56 **(54.82)****
January	8.95 ± 0.21	11.58 ± 1.44 **(33.08)****	7.52 ± 0.63 **(17.95)***	4.64 ± 0.37 **(41.63)****	4.07 ± 0.11 **(48.81)****	3.97 ± 0.34 **(55.85)****
February	10.52 ± 1.86	15.61 ± 0.87 **(48.38)****	9.71 ± 1.01 **(7.69)**	6.79 ± 0.34 **(35.45)****	6.39 ± 0.73 **(39.26)****	5.83 ± 3.43 **(44.58)****

TABLE 61

Photosynthetic Rate of (μ mole CO_2 m^{-2} s^{-1}) the leaves of *C. dichotoma* Growing on Different Concentrations of Fly Ash Amended soil during Different Stages of Growth

	Control	*5%*	*10%*	*20%*	*25%*	*30%*
STAGE 1	19.91 ± 1.43	27.74 ± 2.26 **(39.29)****	16.92 ± 1.62 **(15.03)**	13.58 ± 1.65 **(31.78)****	12.61 ± 0.72 **(36.66)****	11.65 ± 0.95 **(41.47)****
STAGE 2	14.54 ± 1.11	22.37 ± 1.29 **(53.85)****	14.40 ± 1.52 **(0.98)**	9.43 ± 0.96 **(35.11)****	8.27 ± 0.56 **(43.09)****	7.58 ± 0.57 **(47.85)****
STAGE 3	9.65 ± 1.06	13.55 ± 1.08 **(40.47)****	8.74 ± 0.76 **(9.40)**	6.14 ± 0.59 **(36.35)****	5.53 ± 0.56 **(42.64)****	4.85 ± 0.55 **(40.74)****

TABLE 62

Photosynthetic Rate (μ mole CO_2 m^{-2} s^{-1}) of the Leaves of *M. alba* Growing on Different Concentrations of Fly Ash Amended Soil During an Annual Cycle of Three Consecutive Years

	Control	*5%*	*10%*	*20%*	*25%*	*30%*
March	5.23 ± 0.18	5.36 ± 0.47 **(2.49)**	5.39 ± 0.66 **(3.05)**	4.23 ± 0.16 **(19.12)***	3.92 ± 0.41 **(25.05)***	3.26 ± 0.18 **(35.76)****
April	8.05 ± 0.38	8.99 ± 0.38 **(11.68)**	10.72 ± 0.21 **(33.16)****	6.06 ± 0.36 **(24.72)***	5.23 ± 0.34 **(35.03)****	4.05 ± 0.31 **(49.69)****
May	6.62 ± 0.41	7.47 ± 0.22 **(12.84)**	7.71 ± 0.77 **(16.64)***	6.27 ± 0.22 **(5.29)**	5.21 ± 0.51 **(21.29)***	4.37 ± 0.29 **(33.98)****
June	6.24 ± 0.59	7.36 ± 0.17 **(17.95)***	8.61 ± 0.43 **(37.98)****	5.36 ± 0.28 **(14.12)***	5.04 ± 0.23 **(19.23)***	4.15 ± 0.09 **(33.49)****
July	6.21 ± 0.18	6.91 ± 0.32 **(11.27)**	7.91 ± 0.67 **(27.38)****	5.22 ± 0.21 **(15.94)***	5.02 ± 0.12 **(19.16)***	4.24 ± 0.18 **(31.72)****
August	6.17 ± 0.35	6.73 ± 0.32 **(9.07)**	7.42 ± 0.35 **(20.26)***	5.25 ± 0.18 **(14.91)**	4.41 ± 0.23 **(28.53)****	3.42 ± 0.32 **(44.57)****
September	6.09 ± 0.12	6.68 ± 0.55 **(9..69)**	7.51 ± 0.35 **(23.32)****	5.16 ± 0.24 **(15.27)**	4.46 ± 0.27 **(26.77)****	3.35 ± 0.23 **(44.9)****
October	5.59 ± 0.11	5.86 ± 0.73 **(4.83)**	7.21 ± 0.18 **(28.98)****	4.93 ± 0.34 **(11.81)**	4.41 ± 0.33 **(21.11)****	3.21 ± 0.39 **(42.57)****
November	5.03 ± 0.57	5.35 ± 0.28 **(6.36)**	7.07 ± 0.43 **(40.56)****	4.76 ± 0.26 **(5.37)**	3.95 ± 0.05 **(21.47)****	2.87 ± 0.11 **(42.94)****
December	3.97 ± 0.41	4.58 ± 0.48 **(15.37)***	5.19 ± 0.32 **(30.73)****	3.86 ± 0.14 **(2.77)**	2.56 ± 0.43 **835.52)****	2.01 ± 0.05 **(49.87)****
January	3.88 ± 0.47	4.37 ± 0.32 **(12.63)**	4.55 ± 0.51 **(17.29)***	2.94 ± 0.81 **(24.23)***	2.36 ± 0.09 **(39.17)****	1.78 ± 0.06 **(54.12)****
February	4.24 ± 0.32	5.19 ± 0.23 **(2847)****	5.21 ± 0.43 **(28.96)****	3.92 ± 0.34 **(2.97)**	3.06 ± 0.09 **(24.26)**	2.62 ± 0.06 **(35.15)****

TABLE 63

Photosynthetic Rate (μ mole CO_2 m^{-2} s^{-1}) of the leaves of *M. alba* Growing on Different Concentrations of Fly Ash Amended Soil during Different Stages of Growth

	Control	*5%*	*10%*	*20%*	*25%*	*30%*
STAGE 1	6.53 ± 0.39	7.29 ± 0.31 **(11.63)**	8.10 ± 0.36 **(24.06)***	5.48 ± 0.25 **(16.14)***	4.85 ± 0.39 **(25.78)****	3.98 ± 0.21 **(39.05)****
STAGE 2	6.01 ± 0.19	6.54 ± 0.48 **(8.81)**	7.51 ± 0.38 **(24.89)***	5.14 ± 0.24 **(14.55)**	4.57 ± 0.23 **(23.94)***	3.55 ± 0.28 **(40.89)****
STAGE 3	4.23 ± 0.44	4.87 ± 0.32 **(15.19)**	5.50 ± 0.42 **(30.14)****	3.87 ± 0.23 **(8.51)**	2.98 ± 0.16 **(29.49)****	2.31 ± 0.07 **(45.27)****

TABLE 64

Stomatal conductance (μ mole m^{-2} s^{-1}) of the Leaves of *C. dichotoma* Growing on Different Concentrations of Fly Ash Amended Soil During an Annual Cycle of Three Consecutive Years

	Control	*5%*	*10%*	*20%*	*25%*	*30%*
March	0.83 ± 0.05	1.23 ± 0.21 **(48.19)****	0.85 ± 0.07 **(2.41)**	0.81 ± 0.04 **(2.41)**	0.69 ± 0.07 **(16.87)***	0.66 ± 0.08 **(20.49)***
April	1.07 ± 0.07	1.53 ± 0.11 **(42.99)****	1.01 ± 0.09 **(5.61)**	0.82 ± 0.09 **(23.37)***	0.78 ± 0.06 **(27.11)***	0.69 ± 0.05 **(35.52)****
May	0.94 ± 0.07	1.44 ± 0.25 **(53.19)****	0.89 ± 0.07 **(5.32)**	0.82 ± 0.09 **(12.77)**	0.72 ± 0.07 **(23.42)***	0.69 ± 0.05 **(26.59)****
June	0.83 ± 0.07	1.29 ± 0.06 **(55.42)****	0.81 ± 0.04 **(2.41)**	0.73 ± 0.14 **(12.05)**	0.69 ± 0.08 **(16.87)***	0.58 ± 0.11 **(32.53)****
July	0.76 ± 0.05	1.15 ± 0.03 **(5.35)**	0.65 ± 0.09 **(14.47)**	0.64 ± 0.06 **(30.26)****	0.59 ± 0.08 **(35.53)****	0.53 ± 0.07 **(46.05)****
August	0.72 ± 0.06	1.16 ± 0.13 **(84.64)**	0.68 ± 0.08 **(7.69)**	0.66 ± 0.03 **(11.54)**	0.61 ± 0.19 **(21.53)***	0.56 ± 0.05 **(30.76)****
September	0.69 ± 0.09	0.99 ± 0.10 **(58.82)****	0.63 ± 0.05 **(11.76)**	0.61 ± 0.08 **(15.69)***	0.53 ± 0.06 **(31.37)****	0.49 ± 0.07 **(39.22)****
October	0.57 ± 0.02	0.85 ± 0.07 **(57.14)****	0.48 ± 0.02 **(18.37)***	0.47 ± 0.07 **(20.41)***	0.39 ± 0.09 **(36.73)****	0.36 ± 0.08 **(42.86)****
November	0.51 ± 0.05	0.79 ± 0.09 **(62.69)****	0.44 ± 0.07 **(17.07)***	0.38 ± 0.02 **(31.71)****	0.36 ± 0.04 **(36.58)****	0.33 ± 0.01 **(43.91)****
December	0.37 ± 0.04	0.52 ± 0.03 **(40.74)****	0.38 ± 0.06 **(3.71)**	0.31 ± 0.03 **(22.22)***	0.29 ± 0.04 **(29.63)****	0.29 ± 0.02 **(29.63)****
January	0.36 ± 0.05	0.45 ± 0.05 **(45.83)****	0.33 ± 0.02 **(4.17)**	0.29 ± 0.01 **(20.83)***	0.28 ± 0.01 **(25.01)***	0.22 ± 0.02 **(50.49)****
February	0.49 ± 0.04	0.68 ± 0.02 **(47.22)****	0.41 ± 0.04 **(22.46)***	0.39 ± 0.03 **(30.56)****	0.37 ± 0.02 **(41.67)****	0.31 ± 0.01 **(50.11)****

TABLE 65

Stomatal Conductance (μ mole m^{-2} s^{-1}) of the Leaves of *C. dichotoma* Growing on Different Concentrations of Fly Ash Amended Soil during Different Stages of Growth

	Contro'	*5%*	*10%*	*20%*	*25%*	*30%*
STAGE 1	0.91 ± 0.06	1.37 ± 0.15 **(49.59)****	0.87 ± 0.06 **(2.99)**	0.79 ± 0.09 **(13.35)**	0.72 ± 0.69 **(21.53)***	11.65 ± 0.07 **(29.15)***
STAGE 2	0.57 ± 0.05	0.92 ± 0.08 **(61.84)****	0.49 ± 0.06 **(13.16)**	0.45 ± 0.06 **(20.61)****	0.39 ± 0.10 **(31.58)****	29.15 ± 0.34 **(40.35)****
STAGE 3	0.32 ± 0.04	0.51 ± 0.04 **(60.16)****	0.23 ± 0.04 **(13.28)**	0.23 ± 0.0@ **(27.34)****	0.21 ± 0.02 **(34.37)****	0.18 ± 0.01 **(43.75)****

TABLE 66

Stomatal Conductance (μ mole m^{-2} s^{-1}) of the Leaves of *M. alba* Growing on Different Concentrations of Fly Ash Amended Soil during an Annual Cycle of Three Consecutive Years

	Control	*5%*	*10%*	*20%*	*25%*	*30%*
March	0.35 ± 0.01	0.45 ± 0.04 **(28.57)****	0.58 ± 0.03 **(65.71)****	0.31 ± 0.11 **(11.43)**	0.29 ± 0.02 **(*17.14)***	0.28 ± 0.02 **(20.81)***
April	1.03 ± 0.05	1.35 ± 0.07 **(31.07)****	1.67 ± 0.08 **(62.14)****	0.99 ± 0.18 **(3.88)**	0.41 ± 0.03 **(60.19)****	0.33 ± 0.04 **(67.96)****
May	0.85 ± 0.03	1.22 ± 0.03 **(43.53)****	1.58 ± 0.04 **85.88)****	0.49 ± 0.06 **(42.35)****	0.44 ± 0.06 **(48.24)****	0.40 ± 0.01 **(52.94)****
June	0.48 ± 0.05	0.59 ± 0.04 **(22.92)***	0.71 ± 0.06 **(47.92)****	0.37 ± 0.06 **(22.92)**	0.28 ± 0.06 **(41.67)***	0.22 ± 0.03 **(54.17)****
July	0.36 ± 0.04	0.58 ± 0.02 **(61.12)****	0.76 ± 0.07 **(11.43)**	0.35 ± 0.02 **(2.78**	0.23 ± 0.020.02 **(36.46)****	0.13 ± **(63.89)****
August	0.31 ± 0.04	0.51 ± 0.07 **(64.52)****	0.67 ± 0.06 **(116.13)****	0.26 ± 0.01 **(16.13)***	0.21 ± 0.08 **(32.25)****	0.16 ± 0.04 **(48.39)****
September	0.27 ± 0.02	0.47 ± 0.08 **(74.07)****	0.51 ± 0.14 **(88.89)****	0.29 ± 0.02 **(7.41)***	0.18 ± 0.01 **33.07)****	0.14 ± 0.01 **(48.14)****
October	0.25 ± 0.02	0.40 ± 0.05 **(60.43)****	0.45 ± 0.04 **(80.75)****	0.21 ± 0.03 **(16.79)***	0.16 ± 0.02 **(36.11)****	0.11 ± 0.03 **(56.65)****
November	0.24 ± 0.04	0.27 ± 0.04 **(12.05)**	0.31 ± 0.02 **(29.17)****	0.23 ± 0.02 **(4.17)**	0.22 ± 0.07 **(8.33)**	0.18 ± 0.02 **(25.15)***
December	0.14 ± 0.05	0.26 ± 0.05 **(85.71)****	0.26 ± 0.06 **(85.71)****	0.18 ± 0.01 **(28.7)**	0.09 ± 0.02 **(35.71)****	0.08 ± 0.01 **(42.86)****
January	0.13 ± 0.03	0.21 ± 0.05 **(61.54)****	0.23 ± 0.05 **(76.92)****	0.11 ± 0.03 **(22.86)***	0.07 ± 0.03 **(46.15)****	0.08 ± 0.01 **(33.46)****
February	0.23 ± 0.02	0.28 ± 0.04 **(47.06)****	0.28 ± 0.05 **947.06)****	0.18 ± 0.02 **(5.88)**	0.15 ± 0.04 **(17.65)***	0.12 ± 0.02 **(29.41)****

TABLE 67

Stomatal Conductance (μ mole m^{-2} s^{-1}) of the Leaves of *M. alba* Growing on Different Concentrations of Fly Ash Amended Soil During Different Stages of Growth

	Control	*5%*	*10%*	*20%*	*25%*	*30%*
STAGE 1	0.67 ± 0.03	0.90 ± 0.04 **(33.21)****	1.13 ± 0.05 **(67.53)****	0.54 ± 0.10 **(20.29)***	0.35 ± 0.04 **(47.60)****	0.31 ± 0.02 **(54.61)****
STAGE 2	0.29 ± 0.03	0.49 ± 0.05 **(64.71)****	0.59 ± 0.07 **(100.84)****	0.27 ± 0.02 **(6.72)**	0.19 ± 0.03 **(34.45)****	0.13 ± 0.02 **(54.62)****
STAGE 3	0.17 ± 0.03	0.24 ± 0.04 **(45.59)****	0.26 ± 0.04 **(54.41)****	0.18 ± 0.02 **(5.88)**	0.13 ± 0.04 **(23.53)***	0.12 ± 0.01 **(32.35)****

TABLE 68

Intercellular CO_2 Conc. (ppm) of the Leaves of *C. dichotoma* Growing on Different Concentrations of Fly Ash Amended Soil During an Annual Cycle of Three Consecutive Years

	Control	*5%*	*10%*	*20%*	*25%*	*30%*
March	176.43 ± 6.55	148.15 ± 6.17 **(16.03)***	193.43 ± 6.98 **(9.64)**	196.34 ± 6.39 **(11.28)**	258.37 ± 9.12 **(46.44)****	279.68 ± 8.65 **(58.25)****
April	195.22 ± 7.16	156.88 ± 9.57 **(19.64)***	205.53 ± 6.03 **(5.28)**	225.06 ± 8.84 **(15.29)**	261.24 ± 9.65 **(33.86)***	284.54 ± 7.54 **(45.75)****
May	202.19 ± 9.56	162.87 ± 6.73 **(19.,45)***	217.07 ± 8.73 **(7.35)**	223.64 ± 9.92 **(10.61)**	273.43 ± 9.74 **(35.23)****	293.05 ± 8.92 **(44.93)****
June	202.47 ± 7.89	165.36 ± 9.74 **(18.33)***	225.86 ± 9.31 **(11.55)**	224.46 ± 6.74 **(13.58)**	276.17 ± 12.48 **936.41)****	294.52 ± 7.92 **(45.46)****
July	225.77 ± 8.41	171.23 ± 7.24 **(24.16)***	267.57 ± 7.83 **(18.51)***	308.21 ± 7.67 **(36.52)****	236.24 ± 8.38 **(44.51)****	338.59 ± 8.52 **(49.97)****
August	245.67 ± 5.86	195.24 ± 7.12 **(20.53)***	278.76 ± 8.03 **(13.47)**	293.23 ± 14.65 **(19.36)***	335.97 ± 5.57 **(36.76)****	342.94 ± 12.28 **(39.59)****
September	249.68 ± 9.95	183.12 ± 6.69 **(26.65)****	283.85 ± 10.98 **(13.69)**	297.42 ± 8.46 **(19.13)***	338.82 ± 7.14 **(35.71)****	359.06 ± 8.12 **(43.81)****
October	258.64 ± 7.63	176.52 ± 10.63 **(31.75)****	274.37 ± 6.18 **(6.08)**	209.82 ± 8.77 **(19.79)***	346.42 ± 7.84 **(34.09)****	370.14 ± 6.87 **(43.11)****
November	265.38 ± 18.52	199.85 ± 14.65 **(24.69)***	278.24 ± 13.98 **(4.84)**	318.75 ± 14.49 **(20.11)***	345.89 ± 17.86 **(30.34)****	379.41 ± 15.02 **(42.97)****
December	321.41 ± 15.97	258.41 ± 8.85 **(19.61)***	339.79 ± 5.42 **(5.71)**	362.44 ± 10.89 **(12.77)**	375.34 ± 9.13 **(16.78)***	396.27 ± 18.23 **(23.29)***
January	328.41 ± 10.16	289.18 ± 7.19 **(11.95)**	340.68 ± 12.29 **(3.73)**	365.85 ± 8.23 **(11.41)**	379.72 ± 5.64 **(15.62)***	395.65 ± 15.06 **(20.47)***
February	268.22 ± 8.56	212.19 ± 7.98 **(20.89)***	297.34 ± 9.14 **(10.86)**	325.48 ± 12.35 **(21.34)***	353.46 ± 9.26 **(31.76)****	365.41 ± 12.56 **(36.24)****

TABLE 69

Intercellular CO_2 conc. (ppm) of the leaves of *C. dichotoma* Growing on Different Concentrations of Fly Ash Amended Soil During Different Stages of Growth

	Control	*5%*	*10%*	*20%*	*25%*	*30%*
STAGE 1	194.07 ± 7.79	158.31 ± 8.05 **(18.43)***	210.47 ± 7.76 **(8.45)**	218.75 ± 7.97 **(12.71)**	267.30 ± 10.12 **(37.73)****	287.94 ± 8.25 **(48.37)****
STAGE 2	244.94 ± 7.96	181.52 ± 7.92 **(25.89)****	276.13 ± 8.25 **(12.74)**	302.17 ± 9.88 **(23.36)***	336.96 ± 7.23 **(37.57)****	352.68 ± 8.98 **(43.99)****
STAGE 3	295.85 ± 13.30	239.90 ± 9.66 **(18.91)***	314.01 ± 10.27 **(6.13)**	343.13 ± 11.49 **15.98)***	363.59 ± 10.47 **(22.89)***	384.18 ± 15.21 **(29.86)****

TABLE 70

Intercellular CO_2 Conc. (ppm) of the Leaves of *M. alba* Growing on Different Concentrations of Fly Ash Amended Soil During an Annual Cycle of Three Consecutive Years

	Control	*5%*	*10%*	*20%*	*25%*	*30%*
March	259.68 ± 9.94	242.37 ± 29.08 **(6.67)**	205.53 ± 24.72 **(20.85)***	285.82 ± 11.51 **(10.07)**	305.51 ± 27.94 **(17.65)***	326.24 ± 29.63 **(25.63)****
April	238.23 ± 5.28	229.72 ± 12.19 **(3.57)**	177.59 ± 9.45 **(25.45)***	259.17 ± 8.57 **(8.79)**	264.05 ± 8.03 **(10.83)**	275.62 ± 6.78 **(15.69)***
May	249.41 ± 18.05	238.44 ± 79.59 **(4.39)**	187.36 ± 9.74 **(24.87)***	269.69 ± 8.29 **(8.13)**	293.78 ± 6.98 **(17.79)***	320.49 ± 10.28 **(28.49)****
June	275.17 ± 10.86	244.43 ± 20.14 **(11.17)**	221.27 ± 15.77 **(19.59)***	296.69 ± 10.92 **(7.82)**	329.35 ± 27.94 **(19.69)***	320.77 ± 14.15 **(20.21)***
July	280.82 ± 7.05	250.36 ± 6.43 **(10.85)**	230.24 ± 9.62 **(18.01)***	298.24 ± 8.74 **(6.19)**	336.37 ± 11.71 **(19.78)***	348.83 ± 12.84 **(24.21)***
August	309.45 ± 6.45	267.91 ± 10.34 **(13.42)**	232.58 ± 6.69 **(24.84)***	323.18 ± 9.71 **(4.43)**	347.48 ± 8.22 **(12.29)**	357.54 ± 7.52 **(15.54)***
September	316.22 ± 6.02	274.89 ± 21.95 **(13.07)**	241.72 ± 18.19 **(23.56)***	323.44 ± 24.72 **(2.28)**	365.34 ± 11.51 **(15.53)***	385.22 ± 22.53 **(21.82)***
October	316.63 ± 16.73	287.14 ± 7.33 **(9.31)**	261.94 ± 16.93 **(17.27)***	319.52 ± 18.92 **(0.91)**	366.32 ± 22.55 **(15.69)***	385.46 ± 18.07 **(21.74)***
November	336.46 ± 10.72	304.15 ± 25.19 **(9.61)**	286.81 ± 28.03 **(14.76)**	349.81 ± 14.06 **(3.97)**	396.22 ± 18.45 **(1479)**	389.41 ± 13.15 **(15.73)***
December	341.96 ± 9.08	308.78 ± 6.21 **(9.71)**	301.51 ± 5.83 **(11.83)**	355.34 ± 9.67 **(3.91)**	388.49 ± 7.62 **(13.61)**	314.86 ± 23.16 **(15.47)**
January	353.94 ± 20.14	316.09 ± 17.82 **(10.69)**	309.67 ± 12.46 **(12.51)**	369.72 ± 9.65 **(4.46)**	390.66 ± 6.98 **(10.37)**	402.29 ± 12.88 **(13.67)**
February	340.52 ± 19.94	302.19 ± 9.08 **(11.26)**	286.96 ± 16.09 **(15.72)**	346.83 ± 9.93 **(1.85)**	394.85 ± 13.19 **(15.96)***	397.75 ± 18.72 **(16.81)***

TABLE 71

Intercellular CO_2 Conc. (ppm) of the Leaves of *M. alba* Growing on Different Concentrations of Fly Ash Amended Soil During Different Stages of Growth

	Control	*5%*	*10%*	*20%*	*25%*	*30%*
STAGE 1	255.62 ± 10.08	233.74 ± 17.75 **(6.60)**	197.93 ± 14.92 **(22.57)***	277.84 ± 9.82 **(8.69)**	298.17 ± 17.72 **(16.64)***	313.28 ± 15.21 **(22.55)***
STAGE 2	305.78 ± 9.98	270.07 ± 11.51 **(11.67)**	241.62 ± 12.85 **(20.98)***	316.09 ± 15.52 **(3.37)**	352.87 ± 13.49 **(15.72)***	369.26 ± 15.24 **(20.75)***
STAGE 3	343.22 ± 15.01	307.80 ± 14.30 **(10.32)**	296.23 ± 15.60 **(3.69)**	355.42 ± 10.82 **(3.56)**	390.05 ± 11.56 **(13.65)**	396.07 ± 16.97 **(15.40)**

TABLE 72

Stomatal Index of Abaxial Epidermis of the Leaves of *C. dichotoma* Growing on Different Concentrations of Fly Ash Amended Soil During an Annual Cycle of Three Consecutive Years

	Control	5%	10%	20%	25%	30%
March	26.79 ± 3.72	28.66 ± 3.37 **(6.98)**	27.25 ± 2.34 **(1.72)**	26.94 ± 2.93 **(0.56)**	23.52 ± 2.02 **(10.67)**	24.25 ± 2.34 **(8.47)**
April	31.56 ± 3.91	35.51 ± 6.37 **(12.52)**	35.26 ± 4.98 **(11.72)**	34.29 ± 4.25 **(8.65)**	28.31 ± 5.84 **(10.29)**	26.79 ± 3.13 **(15.11)**
May	28.83 ± 4.57	29.92 ± 3.46 **(3.81)**	29.46 ± 3.58 **(2.18)**	27.58 ± 2.88 **(4.34)**	26.53 ± 2.54 **(7.97)**	24.54 ± 1.25 **(14.94)**
June	26.65 ± 3.24	29.68 ± 2.23 **(11.36)**	28.19 ± 2.07 **(5.78)**	28.15 ± 2.98 **(5.62)**	25.99 ± 2.63 **(2.47)**	23.73 ± 2.46 **(10.96)**
July	26.72 ± 2.42	30.93 ± 3.17 **(15.75)**	28.81 ± 1.15 **5.95)**	29.73 ± 3.51 **(11.26)**	21.24 ± 2.23 **(20.58)***	20.67 ± 2.15 **(22.67)***
August	23.86 ± 2.61	26.68 ± 2.19 **(11.82)**	25.36 ± 1.07 **(6.28)**	25.28 ± 1.26 **(5.95)**	22.11 ± 1.95 **(7.34)**	21.37 ± 2.28 **(10.43)**
September	27.84 ± 2.53	30.95 ± 3.79 **(11.17)**	29.16 ± 2.32 **(4.74)**	27.93 ± 2.77 **(0.32)**	24.53 ± 2.62 **(11.89)**	17.66 ± 1.73 **(36.57)****
October	24.83 ± 2.3	32.93 ± 2.92 **(32.62)****	29.46 ± 2.63 **(18.65)***	26.58 ± 3.45 **(7.05)**	24.27 ± 21.89 **(2.26)**	23.52 ± 4.14 **(5.28)**
November	22.21 ± 6.23	26.43 ± 3.04 **(19.05)***	25.06 ± 3.45 **(12.83)**	25.55 ± 2.92 **(15.03)**	22.12 ± 1.26 **(0.41)**	20.84 ± 2.74 **(6.16)**
December	23.58 ± 4.49	25.67 ± 4.81 **(8.78)**	24.65 ± 4.13 **(4.53)**	23.21 ± 2.26 **(1.57)**	16.43 ± 3.34 **(30.32)****	18.44 ± 4.74 **(21.79)****
January	24.38 ± 4.39	36.64 ± 3.48 **(25.67)***	28.16 ± 2.26 **(15.54)***	27.53 ± 6.12 **(12.92)**	22. ± 3.21 **(9.27)**	20.93 ± 2.74 **(14.15)**
February	28.44 ± 3.65	31.53 ± 3.08 **(10.87)**	30.23 ± 3.62 **(6.29)**	28.85 ± 4.91 **(1.44)**	27.32 ± 3.51 **(3.94)**	24.25 ± 3.17 **(14.73)**

TABLE 73

Stomatal Index of Abaxial Epidermis of the Leaves of *C. dichotoma* Growing on Different Concentrations of Fly Ash Amended Soil During Different Stages of Growth

	Control	*5%*	*10%*	*20%*	*25%*	*30%*
STAGE 1	28.45 ± 3.90	30.94 ± 3.85 **(8.74)**	30.04 ± 3.24 **(5.56)**	29.24 ± 3.26 **(2.74)**	26.19 ± 3.25 **(7.96)**	24.89 ± 2.29 **(12.53)**
STAGE 2	25.81 ± 2.46	30.37 ± 3.01 **(17.66)***	28.07 ± 1.79 **(8.75)**	27.38 ± 2.74 **(6.07)**	23.03 ± 2.98 **(10.75)**	20.80 ± 2.57 **(19.39)***
STAGE 3	24.65 ± 4.69	28.56 ± 3.60 **(15.86)***	27.02 ± 3.49 **(9.62)**	26.28 ± 4.05 **(6.62)**	21.99 ± 3.43 **(10.76)**	21.11 ± 3.25 **(14.35)**

TABLE 74

Stomatal Index of Abaxial Epidermis of the Leaves of *M. alba* Growing on Different Concentrations of Fly Ash Amended Soil During an Annual Cycle of Three Consecutive Years

	Control	*5%*	*10%*	*20%*	*25%*	*30%*
March	22.28 ± 1.56	23.41 ± 1.64 **(5.07)**	24.59 ± 2.84 **(10.37)**	20.23 ± 0.66 **(9.21)**	20.18 ± 1.48 **(9.42)**	18.67 ± 0.94 **(16.21)***
April	23.19 ± 2.14	25.12 ± 1.32 **(8.33)**	26.82 ± 2.39 **(15.69)***	22.41 ± 1.62 **(3.36)**	21.59 ± 1.91 **(6.89)**	19.53 ± 0.67 **(15.78)***
May	22.73 ± 2.24	25.88 ± 1.96 **(6.73)**	27.43 ± 1.08 **913.85)**	22.94 ± 2.45 **(0.93)**	20.04 ± 1.14 **(11.83)**	20.96 ± 1.38 **(1.78)**
June	22.56 ± 1.31	15.15 ± 2.61 **(11.48)**	26.07 ± .147 **(15.56)***	21.99 ± 1.62 **(2.53)**	21.74 ± 2.08 **(3.63)**	20.68 ± 1.55 **(8.33)**
July	21.04 ± 2.45	24.53 ± 0.46 **(126.59)***	26.95 ± 2.96 **(28.09)****	21.16 ± 2.91 **(0.57)**	20.01 ± 1.95 **(4.89)**	18.79 ± 2.99 **(10.69)**
August	23.25 ± 1.69	24.12 ± 2.73 **(3.74)**	25.13 ± 1.92 **(8.08)**	21.29 ± 1.23 **(8.43)**	14.41 ± 2.06 **(16.52)***	19.46 ± 2.23 **(16.31)***
September	22.81 ± 0.63	23.38 ± 2.13 **(2.49)**	24.36 ± 4.11 **(6.79)**	20.29 ± 2.71 **(11.04)**	19.66 ± 2.53 **(13.81)**	19.62 ± 4.64 **(13.98)**
October	21.68 ± 1.71 **(3.18)**	22.37 ± 1.85 **(17.98)***	25.59 ± 2.45	20.02 ± 2.55 **(7.65)**	19.86 ± 3.18 **(8.39)**	19.27 ± 1.13 **(11.12)**
November	21.94 ± 1.78	22.92 ± 3.43 **(4.47)**	24.25 ± 2.25 **(20.53)**	20.01 ± 2.41 **98.79)**	19.92 ± 2.08 **(9.16)**	18.52 ± 2.55 **(15.58)***
December	22.43 ± 1.34	23.71 ± 2.17 **(5.71)**	24.62 ± 3.49 **(9.76)**	20.63 ± 0.89 **(8.02)**	18.22 ± 0.52 **(18.76)***	18.17 ± 1.32 **(17.21)***
January	21.72 ± 6.35	22.43 ± 4.82 **(3.26)**	23.71 ± 5.87 **(9.16)**	19.42 ± 5.54 **(10.58)**	19.15 ± 3.91 **(11.83)**	18.47 ± 1.26 **(14.96)**
February	22.65 ± 1.01	22.46 ± 2.94 **(0.83)**	24.94 ± 2.53 **(10.02)**	19.87 ± 1.54 **(12.27)**	18.48 ± (2.18) **(18.41)***	(18.65 ± 0.63 **(17.66)***

TABLE 75

Stomatal Index of Abaxial Epidermis of the leaves of *M. alba* Growing on Different Concentrations of Fly Ash Amended soil during Different Stages of Growth

	Control	*5%*	*10%*	*20%*	*25%*	*30%*
STAGE 1	22.59 ± 1.81	24.89 ± 1.88 **(9.69)**	26.23 ± 1.95 **(15.60)***	21.89 ± 1.58 **(3.51)**	20.88 ± 1.67 **(7.94)**	19.96 ± 1.14 **(12.03)**
STAGE 2	22.19 ± 1.62	23.6 ± 1.79 **(6.33)**	25.50 ± 2.86 **(14.91)**	20.60 ± 2.35 **(6.78)**	19.72 ± 2.43 **(11.08)**	19.28 ± 2.74 **(13.11)**
STAGE 3	22.18 ± 2.62	22.88 ± 3.34 **(3.13)**	24.37 ± 3.53 **(9.87)**	19.98 ± 2.59 **(9.93)**	18.94 ± 2.17 **(14.60)**	18.55) ± 1.44 **(16.37)***

TABLE 76

Stomatal Index of Adaxial Epidermis of the Leaves of *C. dichotoma* Growing on Different Concentrations of Fly Ash Amended Soil During an Annual Cycle of Three Consecutive Years

	Control	*5%*	*10%*	*20%*	*25%*	*30%*
March	6.62 ± 0.53	7.81 ± 0.61 **(17.97)***	7.38 ± 1.07 **(11.48)**	5.26 ± 0.55 **(20.54)***	5.12 ± 0.41 **(22.65)***	4.88 ± 0.25 **(26.28)****
April	5.46 ± 0.74	6.53 ± 0.42 **(19.59)***	5.45 ± 0.91 **(0.18)**	4.47 ± 0.34 **(18.13)***	4.15 ± 0.65 **(23.99)***	4.56 ± 0.08 **(16.48)***
May	7.86 ± 0.36	7.96 ± 0.83 **(1.27)**	7.61 ± 0.45 **(3.180**	6.35 ± 0.58 **(19.21)***	5.53 ± 0.27 **(29.64)****	5.41 ± 0.16 **(31.17)****
June	5.79 ± 0.43	7.69 ± 0.57 **(32.81)****	4.51 ± 0.82 **(22.17)***	4.97 ± 0.62 **(14.16)**	4.55 ± 0.28 **(21.42)***	3.36 ± 0.15 **(41.96)****
July	5.68 ± 0.51	6.37 ± 0.47 **(12.15)**	6.02 ± 0.39 **(5.98)**	5.41 ± 0.58 **(4.75)**	4.34 ± 0.32 **(23.59)***	4.27 ± 0.28 **(24.82)***
August	6.83 ± 1.03	7.64 ± 0.27 **(11.85)**	7.29 ± 0.31 **(6.73)**	6.51 ± 0.46 **94.68)**	5.96 ± 0.18 **(12.73)**	3.39 ± 0.09 **(50.37)****
September	4.37 ± 0.91	5.78 ± 0.59 **(32.27)****	5.54 ± 0.38 **(3.89)**	3.02 ± 0.24 **(30.89)****	3.55 ± 0.46 **(18.76)***	2.28 ± 0.42 **(47.82)****
October	3.86 ± 0.29	4.96 ± 0.71 **(28.49)****	4.61 ± 0.45 **(19.43)***	3.35 ± 0.27 **(13.21)**	2.74 ± 0.52 **(29.01)****	2.12 ± 0.35 **(45.07)****
November	4.96 ± 0.82	6.74 ± 0.36 **(35.87)****	5.12 ± 0.54 **(3.27)**	4.09 ± 0.73 **(17.54)***	4.42 ± 0.64 **(10.89)**	2.82 ± 0.65 **(43.15)****
December	4.52 ± 0.36	5.58 ± 0.43 **(23.45)***	4.38 ± 0.18 **(3.09)**	3.09 ± 0.56 **(31.64)****	2.74 ± 0.26 **(39.38)****	2.26 ± 0.37 **(50)****
January	3.17 ± 0.41	3.84 ± 0.45 **(21.14)***	2.92 ± 0.78 **(7.87)**	1.85 ± 0.38 **(41.64)****	1.61 ± 0.39 **(49.21)****	1.46 ± 0.69 **(53.94)****
February	3.64 ± 0.39	4.72 ± 0.74 **(29.94)****	4.53 ± 0.45 **(24.45)***	2.17 ± 0.27 **(40.38)****	2.42 ± 0.09 **(33.24)****	1.37 ± 0.32 **(62.36)****

TABLE 77

Stomatal Index of Adaxial Epidermis of the Leaves of *C. dichotoma* Growing on Different Concentrations of Fly Ash Amended Soil During Different Stages of Growth

	Control	*5%*	*10%*	*20%*	*25%*	*30%*
STAGE 1	6.43 ± 0.52	7.49 ± 0.61 **(16.56)(**	6.24 ± 0.81 **(3.03)**	5.26 ± 0.53 **(18.19)***	4.83 ± 0.40 **(24.79)***	4.55 ± 0.16 **(29.23)****
STAGE 2	5.19 ± 0.69	6.19 ± 0.51 **(19.33)***	5.16 ± 0.38 **(8.29)**	4.57 ± 0.41 **(11.81)**	4.15 ± 0.37 **(20.09)**	3.02 ± 0.29 **(41.85)****
STAGE 3	4.07 ± 0.49	5.22 ± 0.49 **(28.24)****	4.24 ± 0.49 **(4.05)**	2.80 ± 0.49 **(31.25)****	2.80 ± 0.35 **(31.25)****	1.92 ± 0.51 **(51.45)****

TABLE 78

Stomatal Length (μ m) of Abaxial Epidermis of the Leaves of *C. dichotoma* Growing on Different Concentrations of Fly Ash Amended Soil During an Annual Cycle of Three Consecutive Years

	Control	*5%*	*10%*	*20%*	*25%*	*30%*
March	24.94 ± 3.87	29.92 ± 4.38 **(19.97)***	28.39 ± 1.95 **(13.83)**	25.28 ± 1.57 **(1.36)**	20.68 ± 1.25 **(17.08)***	18.36 ± 1.64 **(26.38)****
April	23.59 ± 2.51	30.87 ± 4.42 **(30.86)****	73.79 ± 4.01 **(17.84)***	28.45 ± 3.65 **(20.61)***	18.67 ± 2.38 **(20.85)***	15.58 ± 1.45 **(33.96)****
May	22.62 ± 3/45	26.13 ± 2/44 **(15.47)**	24.44 ± 1/82 **(7.99)**	24.03 ± 3/79 **(6.14)**	15.93 ± 3.48 **(29.61)****	17.16 ± 2.89 **(24.17)***
June	25.57 ± 1.39	30.05 ± 2.93 **(17.52)***	28.26 ± 2.48 **(12.87)**	25.94 ± 3.17 **(1.45)**	20.07 ± 2.45 **(21.51)***	19.36 ± 4.21 **(24.28)***
July	20.21 ± 2.38	25.14 ± 4.27 **(24.39)***	23.87 ± 3.18 **(18.11)***	23.13 ± 1.96 **(14.45)**	17.78 ± 2.56 **(12.02)**	14.29 ± 3.93 **(29.29)****
August	19.62 ± 4.42	25.83 ± 2.79 **(31.65)****	20.58 ± 1.55 **(4.89)**	24.99 ± 3.08 **(27.37)****	15.82 ± 2.48 **(19.36)***	15.67 ± 2.63 **(20.13)***
September	17.72 ± 2.05	23.35 ± 4.29 **(31.69)****	22.49 ± 2.15 **(26.85)****	20.56 ± 2.45 **(15.96)***	14.72 ± 3.02 **(16.97)***	15.33 ± 2.18 **(13.53)**
October	18.32 ± 2.28	24.56 ± 3.01 **(34.06)****	20.76 ± 2.61 **(13.32)**	19.23 ± 3.87 **(4.97)**	18.27 ± 1.36 **(0.27)**	13.29 ± 2.78 **(27.46)****
November	18.02 ± 3.81	22.64 ± 2.86 **(25/63)****	21.76 ± 2.54 **(20.75)***	17.87 ± 1.64 **(0.83)**	14.72 ± 2.99 **(18.31)***	12.37 ± 2.52 **(31.35)****
December	19.29 ± 1.57	24.83 ± 2.39 **(28.72)****	22.42 ± 4.23 **(16.23)***	19.61 ± 3.56 **(1.66)**	15.38 ± 3.48 **(20.27)***	12.04 ± 1.84 **(37.58)****
January	17.84 ± 2.52	20.77 ± 1.64 **(16.42)***	20.56 ± 2.76 **(15.25)**	15.68 ± 4.39 **(12.17)**	12.19 ± 3.01 **(31.67)****	10.39 ± 2.19 **(41.76)****
February	18.56 ± 2.98	22.92 ± 2.52 **(23.49)***	18.79 ± 3.45 **(1.24)**	19.04 ± 2.54 **(2.58)**	15.34 ± 2.84 **(17.46)***	14.36 ± 2.93 **(22.63)***

TABLE 79

Stomatal Length (μ m) of Abaxial Epidermis of the Leaves of *C. dichotoma* Growing on Different Concentrations of Fly Ash Amended Soil During Different Stages of Growth

	Control	*5%*	*10%*	*20%*	*25%*	*30%*
STAGE 1	24.18 ± 2.80	29.24 ± 3.52 **(20.92)***	27.37 ± 2.81 **(13.18)**	25.92 ± 3.04 **(7.19)**	18.83 ± 2.39 **(22.10)***	17.61 ± 2.54 **(27.16)****
STAGE 2	18.97 ± 2.78	24.72 ± 3.59 **(30.31)****	21.92 ± 2.37 **(15.57)***	21.97 ± 2.84 **(15.85)***	16.64 ± 2.35 **(12.24)**	14.64 ± 2.88 **(22.79)***
STAGE 3	18.42 ± 2.72	22.79 ± 2.35 **(23.67)***	20.88 ± 3.24 **(13.32)**	18.05 ± 3.,03 **(2.05)**	14.40 ± 3.08 **(21.(84)***	12.29 ± 2.37 **(33.30)****

TABLE 80

Stomatal length (μ m) of Abaxial Epidermis of the Leaves of *M. alba* Growing on Different Concentrations of Fly Ash Amended Soil During an Annual Cycle of Three Consecutive Years

	Control	*5%*	*10%*	*20%*	*25%*	*30%*
March	12.23 ± 1.76	12.83 ± 1.25 **(4.950**	13.22 ± 0.65 **(8.09)**	10.63 ± 0.68 **(13.08)**	10.14 ± 1.49 **(17.09)***	9.28 ± 1.78 **(24.12)***
April	15.92 ± 1.39	16.96 ± 2.08 **(6.53)**	17.34 ± 1.68 **(8.92)**	13.82 ± 0.34 **(13.19)**	11.39 ± 0.64 **(28.45)****	10.22 ± 1.18 **(35.84)****
May	17.04 ± 0.87	17.89 ± 1.63 **(4.98)**	19.26 ± 2.62 **(13.020**	14.93 ± 1.73 **(12.38)**	12.13 ± 1.42 **(28.81)***	11.93 ± 1.23 **(29.98)****
June	15.88 ± 1.78	16.31 ± 3.86 **(2.71)**	18.74 ± 1.64 **(18.01)***	14.87 ± 1.65 **(6.36)**	11.81 ± 2.37 **(25.62)****	11.36 ± 1.63 **(28.46)****
July	15.96 ± 2.49	16.68 ± 1.26 **(4.51)**	16.82 ± 1.28 **(5.38)**	13.73 ± 1.44 **(13.97)**	11.69 ± 0.93 **(26.75)****	10.29 ± 0.94 **(35.52)****
August	14.82 ± 1.79	17.86 ± 2.36 **(20.51)***	19.98 ± 1.78 **(34.81)****	12.83 ± 1.73 **(13.42)**	10.78 ± 1.32 **(27.26)****	10.15 ± 1.28 **(31.51)****
September	15.54 ± 2.36	16.29 ± 1.18 **(5.23)**	16.38 ± 2.24 **(5.81)**	10.79 ± 1.91 **(30.29)****	10.18 ± 1.51 **(34.24)****	10.98 ± 1.48 **(29.07)****
October	11.38 ± 1.79	12.41 ± 2.62 **(9.75)**	13.92 ± 2.76 **(22.31)***	10.52 ± 2.08 **(7.56)**	10.46 ± 4.58 **(8.08)**	10.48 ± 2.13 **(7.91)**
November	11.82 ± 1.66	14.29 ± 1.64 **(20.89)***	15.88 ± 2.18 **(34.35)****	9.54 ± 1.68 **(19.28)***	9.78 ± 1.84 **(17.26)***	8.09 ± 1.76 **(31.56)****
December	11.41 ± 2.13	13.98 ± 2.45 **(22.52)***	13.36 ± 1.92 **(17.09)***	10.41 ± 1.78 **(8.76)**	8.78 ± 1.43 **(23.04)***	8.27 ± 1.69 **(27.51)***
January	11.83 ± 1.86	14.02 ± 1.73 **(18.51)***	15.88 ± 0.79 **(34.23)****	9.57 ± 1.65 **(19.13)***	8.41 ± 1.72 **(28.95)****	9.24 ± 1.41 **(21.89)***
February	12.54 ± 0.83	14.43 ± 1.45 **(15.07)**	15.84 ± 1.97 **(26.32)***	10.32 ± 1.03 **(17.73)***	9.69 ± 0.73 **(22.72)***	0.34 ± 1.89 **(25.51)****

TABLE 81

Stomatal Length (μ m) of Abaxial Epidermis of the Leaves of *M. alba* Growing on Different Concentrations of Fly Ash amended Soil During Different Stages of Growth

	Control	*5%*	*10%*	*20%*	*25%*	*30%*
STAGE 1	15.26 ± 1.45	15.99 ± 1.95 **(4.78)**	17.14 ± 1.64 **(12.26)**	13.56 ± 1.35 **(11.16)**	11.36 ± 1.48 **(25.54)****	10.69 ± 1.45 **(29.93)****
STAGE 2	14.41 ± 2.11	15.83 ± 1.85 **(9.85)**	16.77 ± 2.01 **(16.41)***	11.96 ± 1.79 **(16.95)***	10.77 ± 1.33 **(25.21)***	10.47 ± 1.70 **(27.30)****
STAGE 3	11.9 ± 1.62	14.18 ± 1.81 **(19.16)***	15.24 ± 1.71 **(28.07)****	9.96 ± 1.52 **(16.30***	9.16 ± 1.43 **(22.98)***	8.73 ± 1.68 **(26.59)****

TABLE 82

Stomatal Length (μ m) of Adaxial Epidermis of the Leaves of *C. dichotoma* Growing on Different Concentrations of Fly Ash Amended Soil during an Annual Cycle of Three Consecutive Years

	Control	*5%*	*10%*	*20%*	*25%*	*30%*
March	23.44 ± 2.21	30.08 ± 3.02 **(28.32)****	22.31 ± 2.89 **(4.82)**	20.96 ± 3.72 **(10.58)**	26.15 ± 3.85 **(11.56)**	20.49 ± 2.31 **(12.58)**
April	25.85 ± 1.58	28.17 ± 2.42 **(8.97)**	22.28 ± 2.84 **(13.81)**	21.55 ± 3.15 **(16.63)***	23.01 ± 2.71 **(10.98)**	19.64 ± 3.42 **(214.02)***
May	24.76 ± 3.84	25.43 ± 3.36 **(2.71)**	19.24 ± 2.97 **(22.29)***	20.92 ± 2.72 **(15.58)***	22.64 ± 1.57 **(8.56)**	17.62 ± 3.54 **(28.83)****
June	30.85 ± 4.03	30.97 ± 2.17 **(0.39)**	26.38 ± 1.66 **(14.49)**	24.49 ± 2.85 **(20.62)***	28.18 ± 4.21 **(8.65)**	20.07 ± 3.62 **(34.94)****
July	26.58 ± 3.28	27.26 ± 4.18 **(2.56)**	20.74 ± 4.63 **(21.97)***	22.25 ± 2.47 **(16.29)***	25.93 ± 1.48 **(2.45)**	18.54 ± 2.39 **(30.25)***
August	25.73 ± 3.41	28.98 ± 2.29 **(12.63)**	22.17 ± 5.23 **(13.83)**	23.26 ± 7.14 **(9.59)**	26.39 ± 3.36 **(2.56)**	20.29 ± 3.61 **(21.14)***
September	22.24 ± 2.24	24.43 ± 4.51 **(9.84)**	20.47 ± 1.86 **(7.95)**	18.31 ± 1.73 **(17.67)***	21.38 ± 2.67 **(3.86)**	15.54 ± 2.13 **(30.13)****
October	25.29 ± 4.95	27.63 ± 3.86 **(9.25)**	21.18 ± 4.23 **(16.25)***	20.47 ± 1.31 **(19.06)***	25.46 ± 2.07 **(0.67)**	18.52 ± 1.58 **(26.76)****
November	24.64 ± 3.07	25.47 ± 4.25 **(3.36)**	22.98 ± 2.85 **(6.74)**	19.76 ± 3.18 **(19.81)***	22.74 ± 1.53 **(7.71)**	16.03 ± 2.37 **(34.94)****
December	21.25 ± 2.78	22.78 ± 4.92 **(7.02)**	20.24 ± 1.25 **(4.75)**	18.23 ± 2.71 **(14.21)**	21.41 ± 2.29 **(0.75)**	17.87 ± 2.15 **(15.91)***
January	23.73 ± 4.24	25.43 ± 3.12 **(7.16)**	21.47 ± 2.57 **(9.52)**	22.09 ± 2.11 **(6.91)**	20.93 ± 3.86 **(11.79)**	17.42 ± 2.18 **(26.59)****
February	21.96 ± 3.79	25.61 ± 2.63 **(11.62)**	20.75 ± 2.33 **(5.51)**	20.21 ± 1.18 **(7.96)**	19.98 ± 2.04 **(9.01)**	15.86 ± 2.49 **(27.78)****

TABLE 83

Stomatal Length (μ m) of Adaxial Epidermis of the Leaves of *C. dichotoma* Growing on Different Concentrations of Fly Ash Amended Soil During Different Stages of Growth

	Control	*5%*	*10%*	*20%*	*25%*	*30%*
STAGE 1	26.22 ± 2.91	28.66 ± 2.74 **(9.29)**	22.55 ± 2.59 **(14.0)**	21.98 ± 3.11 **16.18)***	24.99 ± 2.08 **(4.69)**	19.45 ± 3.22 **(25.81)****
STAGE 2	24.96 ± 3.46	27.07 ± 3.71 **(8.47**	21.14 ± 3.98 **(15.30(15.57)***	21.07 ± 3.16 **(0.68)**	24.79 ± 2.39 **(26.99)***	18.22 ± 2.42
STAGE 3	22.89 ± 3.77	24.82 ± 3.73 **(8.42)**	21.36 ± 2.25 ***6.70)**	20.07 ± 2.29 **(12.32)**	21.26 ± 2.43 **(7.12)**	16.79 ± 2.29 **(26.64)****

TABLE 84

Stomatal Pore Length (μ m) of Abaxial Epidermis of the Leaves of *C. dichotoma* Growing on Different Concentrations of Fly Ash Amended Soil During an Annual Cycle of Three Consecutive Years

	Control	*5%*	*10%*	*20%*	*25%*	*30%*
March	15.83 ± 3.12	2.015 ± 4.09 **(27.29)****	16.24 ± 3.34 **(2.59)**	19.06 ± 2.36 **(20.43)***	10.16 ± 1.33 **(35.82)****	9.25 ± 2.51 **(41.57)****
April	12.46 ± 2.05	19.63 ± 3.65 **(57.54)****	11.95 ± 2.36 **(4.09)**	16.17 ± 3.72 **(29.78)****	10.55 ± 2.11 **(15.32)**	8.42 ± 3.27 **(32.42)****
May	18.77 ± 2.93	23.98 ± 3.15 **(27.75)****	17.05 ± 4.75 **(9.21)**	19.25 ± 4.92 **(2.56)**	14.56 ± 3.43 **(22.42)***	12.38 ± 2.54 **(34.04)****
June	17.01 ± 2.36	24.32 ± 1.46 **(42.97)****	15.73 ± 2.87 **(7.52)**	23.15 ± 3.45 **(36.09)****	14.84 ± 1.86 **(12.76)**	12.36 ± 2.98 **27.34)****
July	16.11 ± 3.14	20.24 ± 3.38 **(25.64)****	18.34 ± 2.54 **(13.84)**	17.12 ± 4.54 **(6.27)**	9.77 ± 2.58 **(39.35)****	10.64 ± 1.47 **(33.95)****
August	12.63 ± 2.39	16.28 ± 3.52 **(28.89)****	10.36 ± 2.38 **(17.97)**	15.52 ± 3.27 **(22.88)***	9.72 ± 1.82 **(22.96)***	7.48 ± 2.39 **(40.78)****
September	12.75 ± 3.09	18.91 ± 4.38 **(48.31)****	10.52 ± 2.84 **(17.41)***	13.64 ± 3.65 **(6.98)**	10.08 ± 1.85 **(20.94)***	8.15 ± 2.16 **(36.08)****
October	13.27 ± 4.99	18.85 ± 4.74 **(42.05)****	13.15 ± 2.42 **(0.94)**	15.29 ± 4.21 **(15.23)**	10.83 ± 3.92 **(18.39)***	10.05 ± 2.04 **(24.27)***
November	14.84 ± 3.67	19.49 ± 2.24 **(31.34)****	14.37 ± 1.48 **(3.17)**	18.23 ± 3.21 **(22.91)***	11.67 ± 2.13 **(21.36)***	8.86 ± 3.29 **(40.29)****
December	11.63 ± 1.69	15.72 ± 6.67 **(35.17)****	10.31 ± 4.72 **(11.35)**	13.46 ± 2.95 **(15.74)***	7.39 ± 1.96 **(36,45)****	6.79 ± 2.38 **(42.34)****
January	12.73 ± 4.43	16.61 ± 2.54 **(30.48)****	13.46 ± 1.34 **(5.73)**	17.21 ± 2.51 **(32.19)****	9.66 ± 2.64 **(24.12)***	7.29 ± 2.14 **(42.73)****
February	10.02 ± 2.14	14.14 ± 3.65 **(41.12)****	10.84 ± 3.54 **(8.18)**	12.98 ± 2.45 **(29.54)****	7.81 ± 2.36 **(22.05)***	6.74 ± 2.49 **(32.73)****

TABLE 85

Stomatal Pore Length (μ m) of Abaxial Epidermis of the Leaves of *C. dichotoma* Growing on Different Concentrations of Fly Ash Amended Soil During Different Stages of Growth

	Control	*5%*	*10%*	*20%*	*25%*	*30%*
STAGE 1	16.02 ± 2.62	22.02 ± 3.09 **(37.47)****	15.24 ± 3.32 **(4.85)**	19.41 ± 3.61 **(21.16)***	12.53 ± 2.18 **(21.78)***	10.61 ± 2.83 **(33.81)****
STAGE 2	13.69 ± 3.41	18.57 ± 4.01 **(35.64)****	13.09 ± 2.55 **(4.35)**	15.39 ± 3.92 **(12.43)**	10.12 ± 2.55 **(26.21)****	9.08 ± 2.01 **(33.67)****
STAGE 3	12.31 ± 2.98	16.49 ± 3.78 **(30.41)****	12.25 ± 2.78 **(0.49)**	15.47 ± 2.27 **(25.74)****	9.13 ± 2.58 **(25.78)****	7.40 ± **(39.86)****

TABLE 86

Stomatal Pore Length (μ m) of Abaxial Epidermis of the Leaves of *M. alba* Growing on Different Concentrations of Fly Ash Amended Soil During an Annual Cycle of Three Consecutive Years

	Control	*5%*	*10%*	*20%*	*25%*	*30%*
March	10.69 ± 5.32	10.89 ± 1.54 **(1.87)**	12.58 ± 1.63 **(17,68)***	9.67 ± 1.82 **(9.54)**	8.59 ± 1.28 **(19.64)***	8.47 ± 1.09 **(20.76)***
April	10.83 ± 2.58	10.78 ± 2.61 **(0.46)**	13.48 ± 1.79 **(24.47)***	8.96 ± 3.29 **(17.27)***	8.81 ± 3.71 **(18.65)***	8.53 ± 1.24 **(21.23)***
May	15.71 ± 3.52	15.72 ± 1.59 **(0.06)**	16.96 ± 1.52 **(7.95)**	13.69 ± 2.56 **(12.85)**	11.52 ± 1.24 **926.61)****	10.52 ± 3.79 **(33.03)****
June	14.65 ± 1.96	13.96 ± 1.68 **(6.07)**	16.92 ± 14.96 **(15.49)**	13.62 ± 1.76 **(7.03)**	10.46 ± 2.08 **(28.61)****	9.84 ± 1.92 **(32.83)****
July	12.37 ± 2.56	12.61 ± 2.04 **(1.94)**	15.64 ± 3.71 **(26.43)**	10.12 ± 1.57 **(18.19)***	8.56 ± 3.15 **(30.53)****	7.35 ± 1.66 **(40.58)****
August	12.83 ± 2.41	12.18 ± 3.55 **(5.06)**	16.63 ± 1.79 **(29.61)****	11.58 ± 2.13 **(9.74)**	9.34 ± 1.87 **(24.86)***	8.59 ± 1.62 **(33.04)****
September	11.53 ± 1.36	11.83 ± 1.57 **(2.61)**	13.95 ± 1.74 **(20.98)***	9.62 ± 1.31 **(16.56)***	9.42 ± 1.62 **(18.31)***	8.74 ± 1.01 **(24.19)***
October	9.33 ± 2.42	9.06 ± 1.76 **(2.89)**	11.08 ± 3.57 **(18.75)***	6.63 ± 0.74 **(28.93)****	5.14 ± 0.71 **(44.91)****	4.37 ± 0.57 **(53.16)****
November	10.78 ± 1.29	10.65 ± 1.44 **(1.21)**	12.67 ± 0.85 **(17.53)***	8.43 ± 0.94 **(21.79)***	6.77 ± 0.92 **(37.19)****	5.63 ± 0.27 **(47.77)****
December	10.55 ± 0.78	11.16 ± 1.52 **(5.78)**	11.74 ± 1.59 **(1.27)**	9.48 ± 0.76 **(10.14)**	6.94 ± 0.83 **(34.21)****	6.56 ± 0.48 **(37.82)****
January	9.92 ± 1.19	10.63 ± 1.47 **(07.74)**	1.58 ± 2.21 **(16.74)***	6.53 ± 0.71 **(34.14)****	5.57 ± 0.29 **(43.85)****	5.38 ± 0.57 **(45.76)****
February	10.86 ± 1.16	10.18 ± 2.14 **(6.26)**	12.86 ± 2.33 **(18.42)***	8.24 ± 1.02 **(24.12)***	7.66 ± 1.65 **(29.46)****	6.47 ± 0.95 **(40.42)****

TABLE 87

Stomatal Pore Length (μ m) of Abaxial Epidermis of the Leaves of *M. alba* Growing on Different Concentrations of Fly Ash Amended Soil During Different Stages of Growth

	Control	*5%*	*10%*	*20%*	*25%*	*30%*
STAGE 1	12.97 ± 2.79	12.78 ± 1.85 **(1.41)**	14.98 ± 2.47 **(15.53)***	11.48 ± 2.35 **(11.45)**	9.84 ± 2.07 **(24.07)***	9.34 ± 1.88 **(27.98)****
STAGE 2	11.51 ± 2.18	11.42 ± 2.23 **(0.82)**	14.32 ± 2.70 **(24.40)***	9.48 ± 1.43 **(17.61)***	8.19 ± 1.83 **(28.87)****	7.26 ± 1.21 **(36.93)****
STAGE 3	10.52 ± 1.10	10.65 ± 1.64 **(1.21)**	12.21 ± 1.46 **(16.01)***	8.17 ± 0.85 **(22.39)***	6.73 ± 0.92 **(36.02)****	6.01 ± 0.56 **(42.91)****

TABLE 88

Stomatal Pore Length (μ m) of Adaxial Epidermis of the Leaves of *C. dichotoma* Growing on Different Concentrations of Fly Ash Amended Soil during an Annual Cycle of Three Consecutive Years

	Control	*5%*	*10%*	*20%*	*25%*	*30%*
March	26.11 ± 3.84	25.51 ± 3.46 **(2.29)**	20.21 ± 2.23 **(22.59)****	22.22 ± 3.07 **(14.89)**	19.27 ± 1.21 **(26.19)****	17.18 ± 2.42 **(34.21)****
April	25.84 ± 1.68	26.46 ± 1.41 **(2.39)**	23.53 ± 2.34 **(8.93)**	19.85 ± 2.53 **(23.18)****	21.89 ± 5.43 **(15.28)**	18.32 ± 3.56 **(29.12)****
May	18.64 ± 4.17	17.56 ± 2.96 **(5.79)**	18.22 ± 1.12 **(2.25)**	15.56 ± 2.46 **(16.52)***	14.43 ± 3.06 **(22.58)***	14.67 ± 2.51 **(21.29)***
June	26.01 ± 2.64	24.62 ± 2.79 **(5.34)**	24.55 ± 3.62 **(5.61)**	22.48 ± 2.92 **(13.57)**	20.39 ± 1.86 **(21.61)***	18.95 ± 2.76 **(27.14)****
July	26.31 ± 2.81	26.34 ± 3.54 **(0.11)**	21.28 ± 23.61 **(10.15)***	20.98 ± 1.29 **(20.28)***	18.63 ± 1.76 **(2921)****	19.59 ± 2.43 **(25.57)****
August	24.92 ± 2.12	22.84 ± 1.91 **(8.35)**	20.17 ± 1.96 **(19.06)***	21.01 ± 3.07 **(15.69)***	16.54 ± 2.55 **(33.62)****	14.32 ± 1.56 **(42.53)****
September	17.27 ± 4.12	18.08 ± 2.86 **(4.69)**	15.33 ± 1.59 **(11.23)**	15.16 ± 1.82 **(12.21)**	14.17 ± 1.68 **(13.32)**	13.12 ± 1.35 **(24.03)***
October	21.54 ± 2.46	19.28 ± 3.32 **(10.49)**	17.41 ± 2.35 **(19.17)***	16.32 ± 1.78 **(24.23)***	15.69 ± 2.37 **(27.16)****	13.83 ± 2.32 **(35.79)****
November	19.07 ± 3.72	19.76 ± 2.78 **(3.61)**	16.92 ± 2.58 **(11.27)**	15.62 ± 2.13 **(18.09)***	13.54 ± 1.39 **(28.99)****	14.61 ± 1.77 **(23.38)***
December	18.86 ± 2.75	15.37 ± 2.14 **(18.54)***	16.54 ± 1.99 **(12.31)**	14.26 ± 2.64 **(24.39)***	12.19 ± 1.23 **(35.36)****	10.01 ± 2.46 **(46.92)****
January	20.43 ± 1.77	19.76 ± 1.75 **(3.27)**	18.07 ± 2.51 **(11.55)**	15.24 ± 1.84 **(25.41)****	12.22 ± 1.76 **(40.18)****	12.18 ± 1.84 **(40.38)****
February	21.73 ± 2.98	21.84 ± 1.84 **(0.51)**	17.93 ± 2.65 **(17.49)***	16.55 ± 2.06 **(23.83)***	13.25 ± 1.93 **(39.02)****	12.96 ± 2.34 **(40.35)****

TABLE 89

Stomatal Pore Length (μ m) of Adaxial Epidermis of the Leaves of *C. dichotoma* Growing on Different Concentrations of Fly Ash Amended Soil During Different Stages of Growth

	Control	*5%*	*10%*	*20%*	*25%*	*30%*
STAGE 1	24.15 ± 3.08	23.53 ± 2.65 **(2.54)**	21.65 ± 2.32 **(10.44)**	20.02 ± 2.74 **(17.07)***	18.99 ± 2.89 **(21.34)***	17.28 ± 2.81 **(28.44)****
STAGE 2	22.51 ± 2.87	21.63 ± 2.97 **(3.88)**	18.54 ± 2.34 **(17.61)***	18.36 ± 1.99 **(18.41)***	16.45 ± 2.09 **(26.89)****	15.21 ± 1.91 **(32.41)****
STAGE 3	20.02 ± 2.80	19.18 ± 2.12 **(4.19)**	17.36 ± 2.43 **(13.27)**	15.41 ± 2.16 **(22.99)***	12.85 ± 1.57 **(36.07)****	12.44 ± 2.10 **(37.86)****

TABLE 90

Stomatal Width (μ m) of Abaxial Epidermis of the Leaves of *C. dichotoma* Growing on Different Concentrations of Fly Ash Amended Soil during an Annual Cycle of Three Consecutive Years

	Control	*5%*	*10%*	*20%*	*25%*	*30%*
March	9.63 ± 2.65	12.74 ± 1.88 **(32.29)****	13.53 ± 1.34 **(40.49)****	10.51 ± 2.41 **(9.14)**	7.34 ± 1.43 **(23.67)***	8.23 ± 0.73 **(14.53)**
April	8.43 ± 1.14	10.86 ± 2.03 **(28.83)****	12.52 ± 2.35 **(48.51)****	10.75 ± 1.36 **(27.52)****	7.09 ± 2.33 **(15.89)***	7.33 ± 0.32 **(13.05)**
May	7.56 ± 2.54	10.6411.17 ± 1.69 **(40.74)****	17.93 ± 2.43 **(47.75)****	5.57 ± 3.84 **(4.89)**	6.48 ± 1.82 **(26.32)****	± 2.67 **(14.28)**
June	10.91 ± 0.68	11.84 ± 1.93 **(8.52)**	12.38 ± 2.54 **(13.47)**	9.86 ± 1.85 **(9.62)**	8.63 ± 2.39 **(29.89)***	8.81 ± 1.58 **(19.25)***
July	8..66 ± 1.45	9.74 ± 1.35 **12.47**	10.33 ± 1.79 **19.28***	9.26 ± 1.57 **(6.93)**	5.64 ± 1.43 **(34.87)***	6.47 ± 1.61 **(25.29)***
August	8.54 ± 2.67	10.67 ± 2.52 **(24.94)***	11.43 ± 2.01 **(33.84)**	9.37 ± 3.27 **(9.72)**	6.21 ± 2.39 **(27.28)****	6.23 ± 1.47 **(27.04)****
September	6.37 ± 1.86	8.93 ± 1.65 **(40.19)****	10.54 ± 2.38 **(65.46)****	8.35 ± 1.86 **(31.08)****	7.32 ± 1.57 **(14.92)**	8.27 ± 2.15 **(29.83)****
October	10.09 ± 2.26	12.64 ± 2.48 **(25.27)***	12.84 ± 3.27 **(27.25)****	11.37 ± 2.12 **(12.68)**	9.26 ± 1.08 **(8.22)**	9.85 ± 1.46 **(2.37)**
November	7.17 ± 1.48	10.43 ± 1.75 **(45.47)****	10.36 ± 2.65 **(44.49)****	8.62 ± 2.59 **(20.22)***	5.75 ± 1.32 **(19.84)***	5.47 ± 1.95 **(23.79)***
December	--8.52 ± 1.74	10.67 ± 2.39 **(25.23)8**	11.45 ± 1.35 **(34.39)****	10.33 ± 1.69 **(21.24)***	7.53 ± 1.08 **(11.62)**	6.49 ± 0.64 **(23.82)***
January	9.29 ± 1.64	12.21 ± 2.16 **(31.43)****	12.94 ± 1.73 **(39..28)****	10.27 ± 1.45 **(10.54)**	5.34 ± 0.82 **(42.52)****	6.83 ± 1.16 **(26.48)****
February	8.42 ± 0.73	12.23 ± 1.85 **(45.25)****	13.28 ± 2.08 **(57.13)****	9.38 ± 2.36 **(11.41)**	5.37 ± 0.55 **(36.23)****	6.54 ± 1.28 **(22.33)***

TABLE 91

Stomatal Width (μ m) of Abaxial Epidermis of the Leaves of *C. dichotoma* Growing on Different Concentrations of Fly Ash Amended Soil During Different Stages of Growth

	Control	*5%*	*10%*	*20%*	*25%*	*30%*
STAGE 1	9.13 ± 1.75	11.52 ± 1.88 **(26.14)****	12.42 ± 2.16 **(35.77)****	9.76 ± 2.36 **(6.89)**	7.16 ± 1.99 **(21.59)***	7.71 ± 1.32 **(15.55)***
STAGE 2	8.41 ± 2.06	10.49 ± 2.01 **(24.71)***	11.28 ± 2.36 **(34.10)***	9.58 ± 2.20 **(13.93)**	7.10 ± 1.61 **(15.54)***	7.70 ± 1.67 **(8.44)**
STAGE 3	8.35 ± 1.39	11.38 ± 2.03 **(36.35)****	11.99 ± 1.95 **(43.65)****	9.65 ± 2.02 **(15.57)***	5.99 ± 0.94 **(28.17)****	6.33 ± 1.25 **(24.16)***

TABLE 92

Stomatal Width (μ m) of Abaxial Epidermis of the Leaves of *M. alba* Growing on Different Concentrations of Fly Ash Amended Soil During an Annual Cycle of Three Consecutive Years

	Control	*5%*	*10%*	*20%*	*25%*	*30%*
March	5.43 ± 0.74	6.86 ± 0.57 **(26.34)****	7.66 ± 0.61 **(41.07)****	6.07 ± 0.65 **(11.78)**	5.92 ± 0.73 **(9.02)**	5.31 ± 0.86 **(2.03)**
April	5.23 ± 1.12	7.78 ± 0.36 **(48.75)****	7.82 ± 1.24 **(49.52)****	6.83 ± 0.56 **(30.59)****	5.68 ± 0.78 **(8.61)**	5.57 ± 0.83 **(6.59)**
May	5.82 ± 0.85	6.75 ± 1.22 **(15.97)***	7.41 ± 1.49 **(27.31)****	6.36 ± 0.92 **(9.27)**	6.02 ± 1.17 **(3.43)**	5.48 ± 0.83 **(5.84)**
June	6.46 ± 0.72	7.58 ± 0.64 **(17.34)***	7.88 ± 0.85 **(21.98)***	7.14 ± 1.08 **(10.53)**	6.58 ± 0.56 **(1.86)**	6.62 ± 0.68 **(2.48)**
July	5.48 ± 0.64	6.89 ± 0.57 **(25.73)***	6.93 ± 1.05 **(26.45)****	6.09 ± 0.96 **(11.13)**	5.68 ± 0.62 **(3.64)**	5.62 ± 0.41 **(2.55)**
August	5.14 ± 0.73	6.85 ± 1.39 **(33.26)****	7.59 ± 0.54 **(46.88)****	5.73 ± 0.91 **(11.47)**	5.57 ± 0.85 **(8.36)**	5.52 ± 0.69 **(7.39)**
September	4.81 ± 1.32	5.82 ± 1.47 **(20.99)***	6.48 ± 0.89 **(34.72)****	5.37 ± 0.64 **(11.64)**	5.74 ± 0.79 **(19.34)***	5.67 ± 0.37 **(17.87)***
October	3.67 ± 0.96	5.96 ± 0.74 **(62.39)****	6.82 ± 1.13 **(85.83)****	5.09 ± 1.63 **(38.69)****	4.32 ± 0.93 **(17.71)***	4.31 ± 1.19 **(17.43)***
November	4.37 ± 0.65	5.65 ± 0.62 **(29.06)****	5.78 ± 0.59 **(32.26)****	5.48 ± 0.67 **(25.41)***	4.74 ± 0.76 **(8.46)**	4.63 ± 0.39 **(5.94)**
December	3.81 ± 0.86	5.62 ± 0.57 **(47.51)****	5.82 ± 1.46 **(52.76)****	4.36 ± 0.32 **(14.43)**	5.68 ± 0.71 **(49.08)****	3.59 ± 0.58 **(5.77)**
January	4.02 ± 0.59	5.85) ± 1.13 **(45.52)****	6.98 ± 0.64 **(73.63)****	5.47 ± 0.59 **(36.06)****	5.46 ± 0.82 **(35.82)****	4.53 ± 0.62 **(12.68)**
February	4.11 ± 0.53	4.76 ± 0.75 **(15.82)***	6.39 ± 0.69 **(55.47)****	5.22 ± 0.7 **(27.07)****	4.69 ± 0.44 **(14.12)**	4.65 ± 0.38 **(13.28)**

TABLE 93

Stomatal Width (μ m) of Abaxial Epidermis of the Leaves of *M. alba* Growing on Different Concentrations of Fly Ash Amended Soil During Different Stages of Growth

	Control	*5%*	*10%*	*20%*	*25%*	*30%*
STAGE 1	5.73 ± 0.85	7.24 ± 0.69 **(26.29)****	7.69 ± 1.05 **(34.13)****	6.60 ± 0.80 **(15.08)**	6.05 ± 0.81 **(5.49)**	5.74 ± 0.8 **(0.21)**
STAGE 2	4.77 ± 0.91	6.38 ± 1.04 **(33.61)****	6.94 ± 0.89 **(45.45)****	5.57 ± 1.03 **(16.65)***	5.32 ± 0.80 **(11.57)**	5.28 ± 0.79 **(10.58)**
STAGE 3	4.07 ± 0.65	5.47 ± 0.77 **(34.15)****	6.24 ± 0.84 **(53.09)****	5.13 ± 0.57 **(25.87)****	5.14 ± 0.68 **(26.11)****	4.35 ± 0.49 **(6.68)**

TABLE 94

Stomatal Width (μ m) of Adaxial Epidermis of the Leaves of *C. dichotoma* Growing on Different Concentrations of Fly Ash Amended Soil During an Annual Cycle of Three Consecutive Years

	Control	*5%*	*10%*	*20%*	*25%*	*30%*
March	9.12 ± 1.86	16.97 ± 1.35 **(86.07)****	16.02 ± 0.89 **(75.65)****	14.72 ± 0.52 **(61.43)****	12.41 ± 0.35 **(36.07)****	10.06 ± 0.72 **(10.31)**
April	10.27 ± 0.46	18.31 ± 1.45 **(78.39)****	16.64 ± 1.67 **(62.02)****	17.56 ± 0.75 **(70.98)****	12.89 ± 0.43 **(25.51)****	9.76 ± 0.74 **(4.96)**
May	9.49 ± 0.75	15.02 ± 0.98 **(58.27)****	14.89 ± 1.02 **(56.92)****	14.17 ± 0.52 **(49.31)****	12.43 ± 0.84 **(30.98)****	11.21 ± 0.98 **(18.12)***
June	11.55 ± 0.69	19.68 ± 1.95 **(70.39)****	18.25 ± 1.41 **(58.08)****	18.13 ± 0.87 **(56.96)****	15.05 ± 0.63 **(30.33)****	12.29 ± 1.02 **(6.41)**
July	8.32 ± 1.62	13.64 ± 1.86 **(63.94)****	12.42 ± 1.66 **(49.27)****	10.23 ± 0.74 **(22.95)***	9.85 ± 0.63 **(18.38)***	9.45 ± 0.57 **(13.58)**
August	13.51 ± 1.24	19.73 ± 0.92 **(46.04)****	15.29 ± 1.42 **(13.17)**	18.34 ± 1.55 **(35.75)****	14.28 ± 1.39 **(5.69)**	13.87 ± 0.52 **(2.66)**
September	9.94 ± 0.83	14.52 ± 2.34 **(46.27)****	12.74 ± 1.69 **(218.17)****	10.93 ± 1.66 **(9.85)**	10.64 ± 1.49 **(7.04)**	8.15 ± 1.51 **(18.01)***
October	7.84 ± 0.54	13.27 ± 2.45 **(69.26)****	13.09 ± 1.95 **(66.96)****	10.48 ± 1.18 **(33.67)****	9.63 ± 1.32 **(22.83)***	0.75 ± 1.68 **(24.36)***
November	8.34 ± 1.74	12.52 ± 1.96 **(50.12)****	11.18 ± 2.81 **(34.05)****	10.64 ± 1.43 **(27.57)****	9.59 ± 0.65 **(14.98)**	9.84 ± 0.96 **(17.98)***
December	8.88 ± 0.92	11.36 ± 0.65 **(217.93)****	12.67 ± 1.84 **(42.68)****	11.53 ± 1.82 **(29.84)****	10.34 ± 0.98 **(16.44)***	9.01 ± 1.28 **(1.47)**
January	9.07 ± 0.63	12.19 ± 0.97 **(23.37)***	10.15 ± 1.17 **(11.97)**	9.42 ± 0.76 **(3.85)**	8.65 ± 1.47 **(4.62)**	8.05 ± 0.76 **(11.25)**
February	8.96 ± 0.56	13.34 ± 1.04 **(48.88)****	11.81 ± 1.74 **(31.81)****	11.27 ± 1.28 **(25.78)****	9.96 ± 1.63 **(11.16)**	9.52 ± 0.82 **6.25)**

TABLE 95

Stomatal Width (μ m) of Adaxial Epidermis of the Leaves of *C. dichotoma* Growing on Different Concentrations of Fly Ash Amended Soil During Different Stages of Growth

	Control	*5%*	*10%*	*20%*	*25%*	*30%*
STAGE 1	10.11 ± 0.94	17.49 ± 1.43 **(73.11)****	16.45 ± 1.24 **(62.75)****	16.14 ± 0.66 **(59.73)****	13.19 ± 0.56 **(30.54)****	10.83 ± 0.86 **(7.14)**
STAGE 2	9.90 ± 1.05	15.29 ± 1.89 **(54.45)****	13.38 ± 1.68 **(35.16)****	12.49 ± 1.28 **(26.16)****	11.15 ± 1.21 **(12.09)**	10.30 ± 1.07 **(4.06)**
STAGE 3	8.81 ± 0.96	12.10 ± 1.15 **(37.33)****	11.45 ± 1.89 **(29.96)****	10.71 ± 1.32 **(21.59)***	9.63 ± 1.18 **(9.33)**	9.10 ± 0.95 **(3.32**

TABLE 96

Stomatal Pore Width (μ m) of Abaxial Epidermis of the Leaves of *C. dichotoma* Growing on Different Concentrations of Fly Ash Amended Soil During an Annual Cycle of Three Consecutive Years

	Control	*5%*	*10%*	*20%*	*25%*	*30%*
March	5.14 ± 0.46	6.98 ± 0.54 **(35.79)****	5.63 ± 0.89 **(9.53)**	3.45 ± 0.39 **(32.87)****	2.44 ± 0.97 **(56.42)****	2.29 ± 0.74 **(55.45)****
April	6.27 ± 0.91	10.43 ± 10.2 **(66.35)****	903 ± 1.27 **(44.02)****	4.94 ± 0.92 **(21.21)***	3.49 ± 0.73 **(44.34)****	2.13 ± 0.35 **(66.03)****
May	6.46 ± 0.74	8.55 ± 0.59 **(2.79)**	6.28 ± 0.46 **(32.36)****	4.17 ± 1.12 **(35.45)****	3.19 ± 0.67 **(39.47)****	3.57 ± 0.93 **(44.73)****
June	7.34 ± 1.88	8.68 ± 1.21 **(18.26)***	7.92 ± 0.82 **(7.91)**	6.05 ± 0.63 **(17.57)***	2.99 ± 0.46 **(59.26)****	3.62 ± 0.51 **(50.68)****
July	4.74 ± 0.52	7.21 ± 0.61 **(52.19)****	5.18 ± 0.36 **(9.28)**	3.03 ± 0.27 **(36.07)****	1.48 ± 0.81 **(68.77)****	0.67 ± 0.42 **(85.86)****
August	5.96 ± 1.16	8.35 ± 0.49 **(40.11)****	6.08 ± 0.98 **(2.01)**	5.79 ± 0.73 **(2.85)**	0.72 ± 0.41 **(54.36)****	1.14 ± 0.35 **(80.87)****
September	5.53 ± 0.98	8.26 ± 1.25 **(49.37)****	7.14 ± 1.09 **(29.11)****	3.76 ± 0.38 **(32.07)****	2.43 ± 0.61 **(56.05)****	1.36 ± 0.29 **(75.46)****
October	3.27 ± 0.84	5.02 ± 0.93 **(53.51)****	4.98 ± 0.75 **(52.29)****	2.16 ± 0.64 **(33.94)****	1.82 ± 0.83 **(44.34)****	1.64 ± 0.59 **(49.85)****
November	3.73 ± 0.27	5.38 ± 0.49 **(44.24)****	3.96 ± 0.36 **(6.17)**	1.84 ± 0.52 **(50.67)****	0.67 ± 0.44 **(82.04)****	0.55 ± 0.43 **(85.26)****
December	4.46 ± 0.57	6.26 ± 0.72 **(40.36)****	6.14 ± 0.83 **(37.67)****	2.73 ± 0.61 **(38.79)****	1.43 ± 0.52 **(67.94)****	1.36 ± 0.61 **(69.51)****
January	3.96 ± 0.22	5.82 ± 0.15 **(6.97)****	4.07 ± 0.68 **(2.78)**	3.67 ± 0.57 **(7.32)**	1.31 ± 0.34 **(66.92)****	1.04 ± 0.28 **(73.74)****
February	4.56 ± 0.28	7.26 ± 0.64 **(56.13)****	5.58 ± 0.34 **(20)***	1.39 ± 0.317 **(70.11)****	0.69 ± 0.29 **(85.16)****	1.27 ± 0.34 **(72.69)****

TABLE 97

Stomatal Pore Width (μ m) of Abaxial Epidermis of the Leaves of *C. dichotoma* Growing on Different Concentrations of Fly Ash Amended Soil During Different Stages of Growth

	Control	*5%*	*10%*	*20%*	*25%*	*30%*
STAGE 1	6.30 ± 0.99	8.68 ± 0.80 **(37.40)****	7.21 ± 0.89 **(14.48)**	4.65 ± 0.76 **(26.18)****	3.15 ± 0.70 **(49.90)****	2.90 ± 0.63 **(53.94)****
STAGE 2	4.37 ± 0.87	7.21 ± 0.82 **(47.89)****	5.84 ± 0.79 **(19.89)***	3.68 ± 0.51 **(24.41)***	2.11 ± 0.66 **(56.67)****	1.20 ± 0.41 **(73.33)****
STAGE 3	4.24 ± 0.33	6.18 ± 0.5 **(47.14)****	4.93 ± 0.55 **(17.56)***	2.40 ± 0.46 **(42.68)****	1.02 ± 0.39 **(75.59)****	1.05 ± 0.41 **(74.88)****

TABLE 98

Stomatal Pore Width (μ m) of Abaxial Epidermis of the Leaves of *M. alba* Growing on Different Concentrations of Fly Ash amended Soil During an Annual cycle of Three Consecutive Years

	Control	*5%*	*10%*	*20%*	*25%*	*30%*
March	2.75 ± 0.41	3.55 ± 0.72 **(29,09)****	0.76 ± 0.67 **(36.73)****	0.18 ± 0.26 **(15.63)***	2.63 ± 0.13 **(4.36)**	2.23 ± 0.36 **(18.91)***
April	3.48 ± 0.78	4.06 ± 0.54 **(16.67)***	4.76 ± 0.28 **(36.78)****	3.52 ± 0.43 **(1.15)**	3.12 ± 0.23 **(10.34)**	2.67 ± 0.28 **(23.27)***
May	3.15 ± 0.52	4.97 ± 0.12 **(25.82)****	5.42 ± 0.78 **(37.21)****	3.78 ± 0.73 **(4.33)**	2.92 ± 0.31 **(26.07)****	2.52 ± 0.26 **(36.22)****
June	4.27 ± 0.91	4.56 ± 0.78 **(7.03)**	5.38 ± 0.23 **(25.99)****	0.37 ± 0.63 **(2.34)**	3.48 ± 0.34 **(18.51)***	3.15 ± 0.23 **(26.22)****
July	3.46 ± 0.71	3.99 ± 0.96 **(15.31)**	4.62 ± 0.42 **(33.52)****	3.51 ± 0.16 **(1.45)**	2.76 ± 0.12 **(20.53)***	2.59 ± 0.25 **(25.77)****
August	3.92 ± 0.56	4.91 ± 0.52 **(25.26)***	5.46 ± 0.31 **(39.28)****	4.72 ± 0.18 **(20.41)***	2.52 ± 0.19 **(35.71)****	1.92 ± 0.36 **(51.02)****
September	3.69 ± 0.12	4.58 ± 0.76 **(24.11)***	5.01 ± 0.12 **(36.85)****	3.93 ± 0.34 **(6.54)**	2.79 ± 0.18 **(24.39)***	2.35 ± 0.53 **(36.04)****
October	2.47 ± 0.62	3.52 ± 0.76 **(42.51)****	4.12 ± 0.93 **(66.81)****	2.68 ± 0.19 **(8.52)**	2.05 ± 0.59 **(17.04)***	1.69 ± 0.38 **(31.57)****
November	2.55 ± 0.57	3.73 ± 0.69 **(46.27)****	4.26 ± 0.35 **(67.05)****	3.51 ± 0.23 **(37.64)****	1.76 ± 0.27 **(30.98)****	1.62 ± 0.07 **(36.47)****
December	2.73 ± 0.41	3.54 ± 0.27 **(29.67)****	3.71 ± 0.51 **(35.89)****	2.98 ± 0.12 **(9.16)**	1.83 ± 0.19 **(32.61)****	1.63 ± 0.28 **(40.29)****
January	2.95 ± 0.67	2.98 ± 0.73 **(1.02)**	4.08 ± 0.31 **(38.31)****	2.56 ± 0.46 **(13.22)**	0.02 ± 0.41 **(31.52)****	1.31 ± 0.35 **(55.59)****
February	3.07 ± 0.27	3.68 ± 0.85 **(19.86)***	4.63 ± 0.43 **(50.81)****	3.37 ± 0.25 **(9.77)**	2.76 ± 0.24 **(10.09)**	1.81 ± 0.11 **(41.04)****

TABLE 99

Stomatal Pore Width (μ m) of Abaxial Epidermis of the Leaves of *M. alba* Growing on Different Concentrations of Fly Ash Amended Soil During Different Stages of Growth

	Control	*5%*	*10%*	*20%*	*25%*	*30%*
STAGE 1	3.61 ± 0.65	4.28 ± 0.54 **(18.68)***	4.83 ± 0.49 **(33.70)****	3.71 ± 0.51 **(2.77)**	3.03 ± 0.25 **(15.92)***	2.64 ± 0.28 **(26.85)****
STAGE 2	3.38 ± 0.50	4.25 ± 0.75 **(25.55)****	4.81 ± 0.44 **(42.09)****	3.71 ± 0.21 **(9.60)**	2.53 ± 0.26 **(25.26)***	2.14 ± 0.38 **(36.78)****
STAGE 3	2.82 ± 0.48	3.48 ± 0.62 **(23.27)***	4.17 ± 0.40 **(47.61)****	3.10 ± 0.26 **(9.91)**	2.09 ± 0.27 **(25.84)****	1.59 ± 0.20 **(43.62)****

TABLE 100

Stomatal Pore Width (μ m) of Adaxial Epidermis of the Leaves of *C. dichotoma* Growing on Different Concentrations of Fly Ash Amended Soil During an Annual Cycle of Three Consecutive Years

	Control	*5%*	*10%*	*20%*	*25%*	*30%*
March	6.41 ± 1.07	5.49 ± 0.58 **(14.35)**	6.25 ± 0.32 **(2.49)**	4.06 ± 0.75 **(36.67)****	4.93 ± 0.35 **(23.08)***	3.97 ± 0.29 **(45.86)****
April	7.59 ± 0.94	6.65 ± 0.215 **(12.38)**	7.28 ± 0.59 **(4.08)**	6.15 ± 0.22 **(18.97)***	4.46 ± 0.41 **(41.23)****	3.98 ± 0.74 **(47.56)****
May	7.26 ± 0.38	7.99 ± 0.61 **(10.05)**	7.61 ± 0.27 **(4.82)**	6.17 ± 0.53 **(15.01(20.93)***	5.74 ± 0.62 **(28.09)****	5.22 ± 0.43
June	5.29 ± 0.43	4.82 ± 0.39 **(8.84)**	4.63 ± 042 **(12.47)**	3.48 ± 0.61 **(34.21)****	3.15 ± 0.83 **(40.45)****	3.47 ± 0.51 **(34.41)****
July	6.41 ± 0.65	5.44 ± 0.72 **(15.31)**	6.27 ± 0.43 **(2.18)**	4.56 ± 0.81 **(28.86)****	3.42 ± 0.54 **(46.64)****	3.26 ± 0.61 **(46.14)****
August	6.16 ± 0.92	6.73 ± 0.81 **(9.25)**	6.96 ± 0.54 **(12.98)**	4.71 ± 0.95 **(23.53)***	5.09 ± 0.56 **(17.37)***	4.28 ± 0.76 **(30.52)****
September	5.22 ± 0.64	4.43 ± 0.37 **(15.13)**	5.06 ± 0.74 **(3.06)**	3.72 ± 0.37 **(28.73)****	2.18 ± 0.42 **(58.23)****	1.44 ± 0.31 **(72.41)****
October	5.24 ± 0.52	5.17 ± 0.28 **(1.34)**	5.20 ± 0.48 **(0.76)**	3.25 ± 0.75 **(37.97)****	2.08 ± 0.21 **(60.31)****	1.89 ± 0.58 **(63.93)****
November	5.78 ± 0.66	4.82 ± 0.62 **(16.61)***	5.63 ± 0.59 **(2.59)**	4.17 ± 0.43 **(27.85)****	4.36 ± 0.27 **(24.56)***	3.97 ± 0.36 **(31.48)****
December	4.29 ± 0.57	3.46 ± 0.29 **(19.35)***	3.22 ± 0.47 **(24.94)***	2.13 ± 0.68 **(50.34)****	1.97 ± 0.48 **(54.07)****	1.25 ± 0.38 **(70.86)****
January	4.01 ± 0.41	3.66 ± 0.53 **(8.72)**	3.93 ± 0.82 **(1.99)**	2.32 ± 0.73 **(42.14)****	2.39 ± 0.32 **(40.39)****	2.07 ± 0.26 **(48.37)****
February	4.63 ± 0.32	3.94 ± 0.43 **(14.91)**	4.23 ± 0.61 **(8.62)**	3.09 ± 0.87 **(33.26)****	2.81 ± 0.48 **(39.38)****	2.52 ± 0.17 **(45.57)****

TABLE 101

Stomatal Pore Width (μ m) of Adaxial Epidermis of the Leaves of *C. dichotoma* Growing on Different Concentrations of Fly Ash Amended Soil During Different Stages of Growth

	Control	*5%*	*10%*	*20%*	*25%*	*30%*
STAGE 1	6.63 ± 0.70	6.23 ± 0.45 **(6.02)**	6.44 ± 0.41 **(2.94)**	4.96 ± 0.52 **(25.19)***	4.57 ± 0.55 **(31.15)****	4.03 ± 0.49 **(39.21)****
STAGE 2	5.75 ± 0.68	5.44 ± 0.54 **(5.47)**	5.87 ± 0.72 **1.99)**	4.06 ± 0.43 **29.48)****	3.19 ± 0.56 **(44.55)****	2.71 ± **(52.80)****
STAGE 3	4.67 ± 0.49	3.97 ± 0.46 **(15.21)**	4.25 ± 0.62 **(9.08)**	2.92 ± 0.67 **(37.41)****	2.88 ± 0.38 **(38.37)****	2.45 ± 0.29 **(47.56)****

TABLE 102

Soil pH at Different Fly ash Concentrations as Against Control Unamended Soil Before and After Sampling for Both the Species

	Control	*5%*	*10%*	*20%*	*25%*	*30%*
SOIL	6.88 ± 0.23	6.92 ± 0.17 **(0.58)**	6.98 ± 0.34 **(1.45)**	7.04 ± 0.09 **(2.33)**	7.06 ± 0.27 **(2.62)**	7.09 ± 0.21 **(3.05)**
SOIL (*C. dichotoma*)	6.69 ± 0.21	6.74 ± 0.07 **(0.75)**	6.75 ± 0.17 **(0.89)**	6.88 ± 0.25 **(0.89)**	6.91 ± 0.16 **(2.84)**	6.95 ± 0.09 **(3.05)**
SOIL (*M. alba*)	6.79 ± 0.34	6.83 ± 0.29 **(0.59)**	6.87 ± 0.31 **(1.18)**	7.01 ± 0.08 **(3.24)**	6.99 ± 0.34 **(2.95)**	7.03 ± 0.09 **(3.53)**

TABLE 103

Soil Sulphur (μ gm g^{-1} dry wt.) at Different Fly Ash Concentrations as Against Control unamended Soil Before and After Sampling for Both the Species

	Contrcl	*5%*	*10%*	*20%*	*25%*	*30%*
SOIL	52.63 ± 2.65	52.98 ± 3.28 **(0.64)**	53.15 ± 3.23 **(0.98)**	54.08 ± 4.11 **(2.76)**	54.17 ± 5.64 **(3.92)**	54.36 ± 3.93 **(3.28)**
SOIL (*C. dichotoma*)	46.85 ± 4.12	44.57 ± 5.51 **(94.87)**	46.52 ± 2.38 **(0.71)**	48.87 ± 4.65 **(4.32)**	50.38 ± 3.31 **(7.53)**	50.94 ± 4.19 **(8.73)**
SOIL (*M. alba*)	48.27 ± 3.45	45.81 ± 3.09 **(5.09)**	44.07 ± 5.12 **(8.71)**	46.17 ± 5.27 **(4.35)**	47.34 ± 3.42 **(1.93)**	49.78 ± 4.88 **(3.12)**

TABLE 104

Soil Nitrogen (mg g^{-1} dry wt.) at Different Fly Ash Concentrations as Against Control Unamended Soil Before and After Sampling for Both the Species

	Control	*5%*	*10%*	*20%*	*25%*	*30%*
SOIL	0.64 ± 0.13	0.68 ± 0.11 **(6.25)**	0.69 ± 0.02 **(7.81)**	0.79 ± 0.34 **(23.42)***	0.82 ± 0.36 **(28.13)****	0.86 ± 0.31 **(34.38)****
SOIL (*C. dichotoma*)	0.58 ± 0.21	0.50 ± 0.17 **(13.79)**	0.54 ± 0.08 **(6.89)**	0.62 ± 0.05 **(6.91)**	0.65 ± 0.24 **(12.07)**	0.73 ± 0.13 **(25.86)****
SOIL (*M. alba*)	0.56 ± 0.15	0.55 ± 0.26 **(1.78)**	0.41 ± 0.07 **(26.78)****	0.55 ± 0.12 **(1.79)**	0.67 ± 0.23 **(19.64)***	0.69 ± 0.07 **(23.12)***

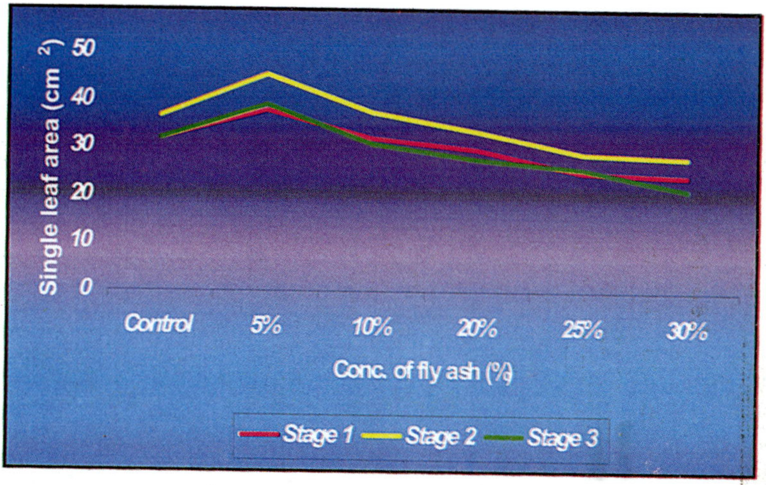

Figure 4: Comparison of single leaf area (cm^2) of *C. dichotoma* during different stages of growth and at various fly ash concentration in soil

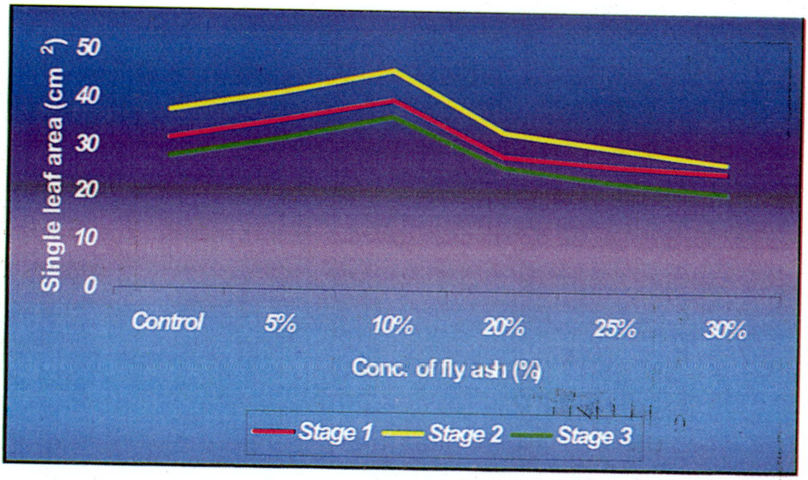

Figure 5: Comparison of single leaf area (cm^2) of *M. alba* during different stages of growth and at various fly ash concentrations in soil

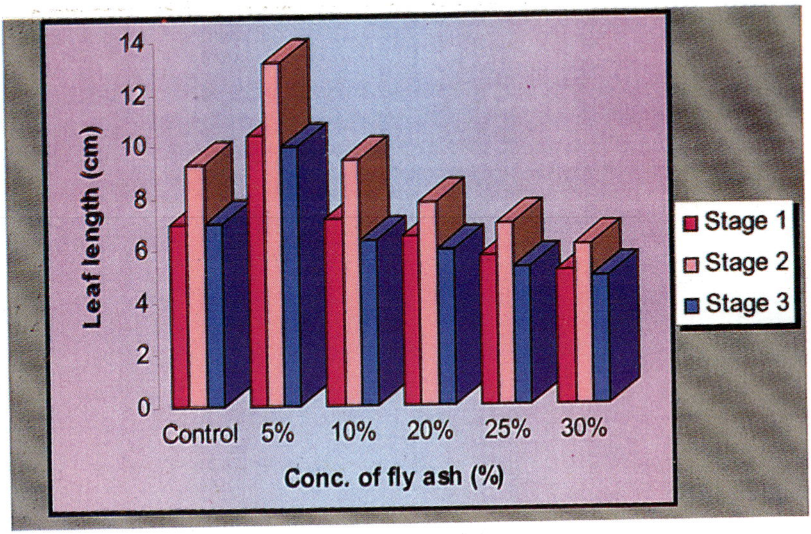

Figure 6: Comparison of leaf length (cm) of *C.dichotoma* during different different stages of growth and at various fly ash concentrations in soil

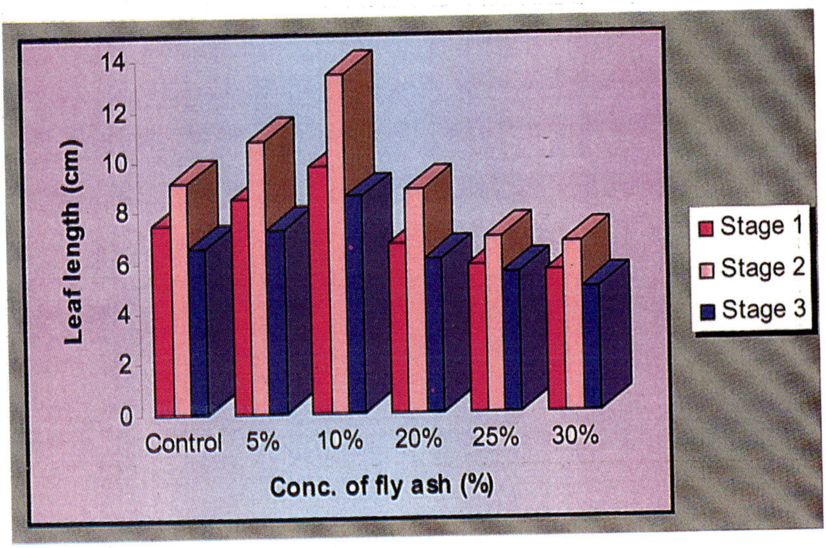

Figure 7: Comparison of leaf length (cm) of *M. alba* during different stages of growth and at various fly ash concentrations in soil.

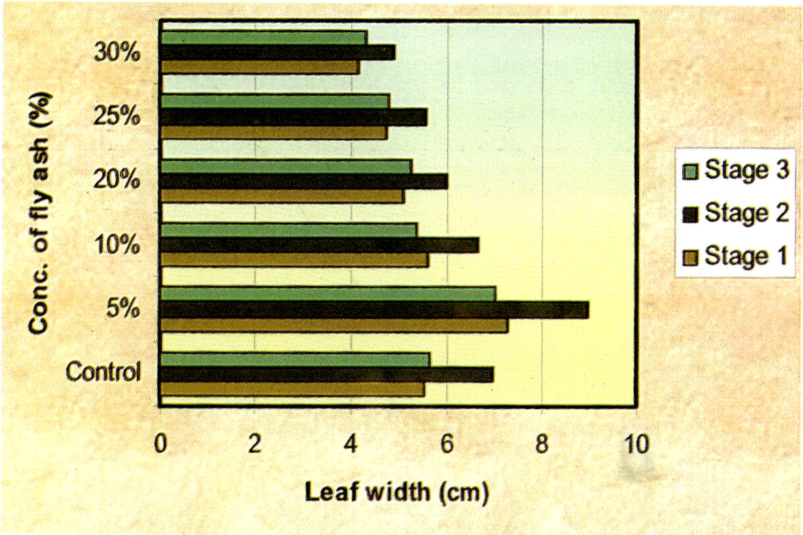

Figure 8: Comparison of leaf width (cm) of *C. dichotoma* during different stages of growth and at various fly ash concentrations in soil.

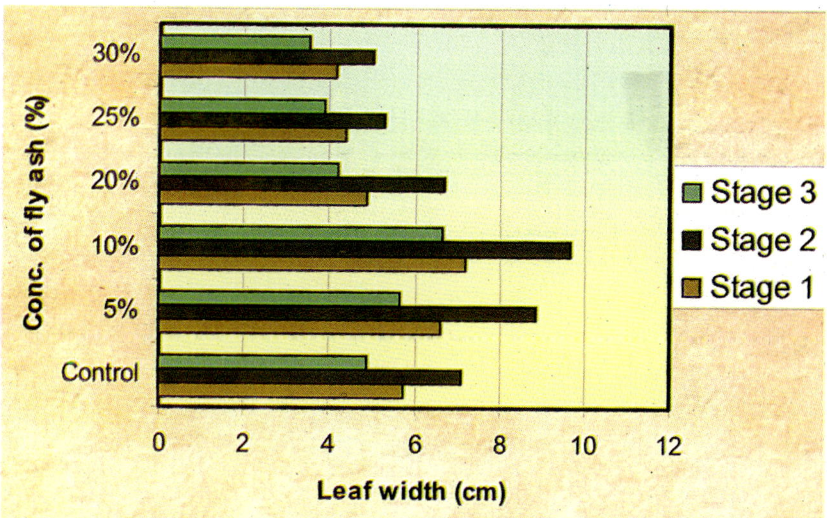

Figure 9: Comparison of leaf width (cm) of *M. alba* during different stages of growth and at various fly ash concentrations in soil

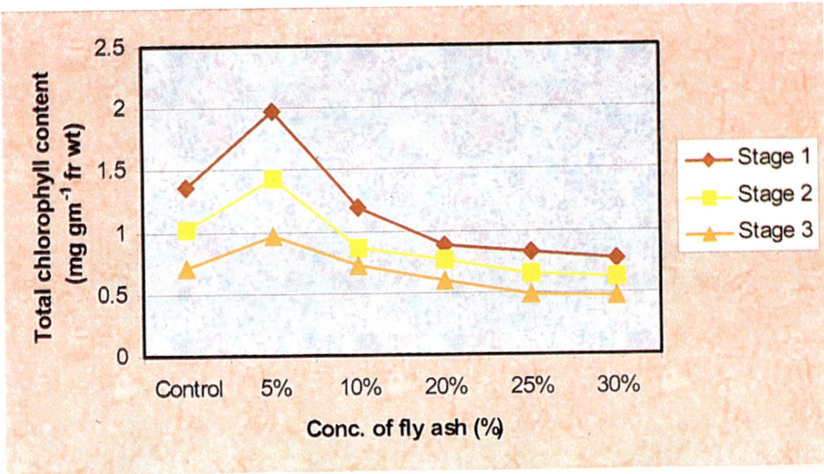

Figure 10: Comparison of chlorophyll (mg gm^{-1} fr. wt.) of the leaves of *C. dichotoma* during different stages of growth and at various fly ash concentrations in soil

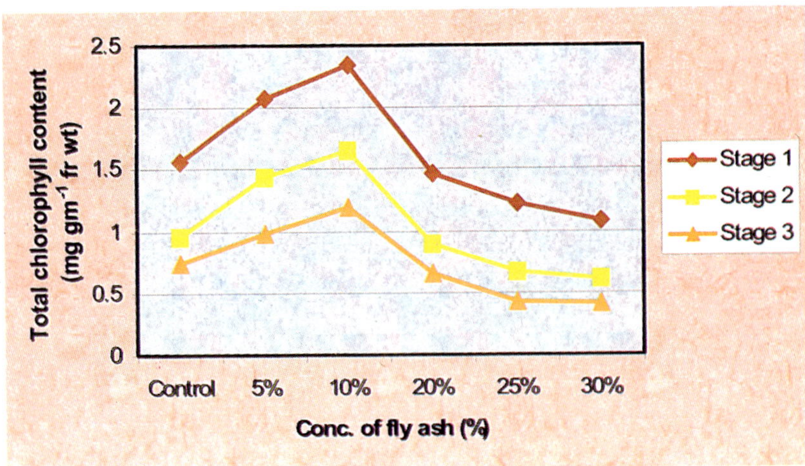

Figure 11: Comparison of total chlorophyll (mg gm^{-1} fr. wt.) of the Leaves of *M. alba* during different stages of growth and at various fly ash concentrations in soil

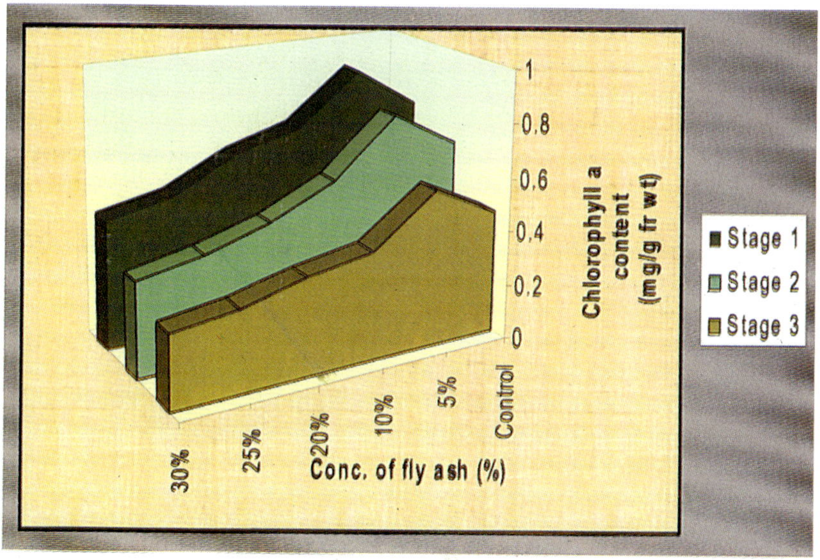

Figure 12: Comparison of chlorophyll a (mg gm^{-1} fr. wt.) of the leaves of *C. dichotoma* during different stages of growth and at various fly ash concentrations in soil

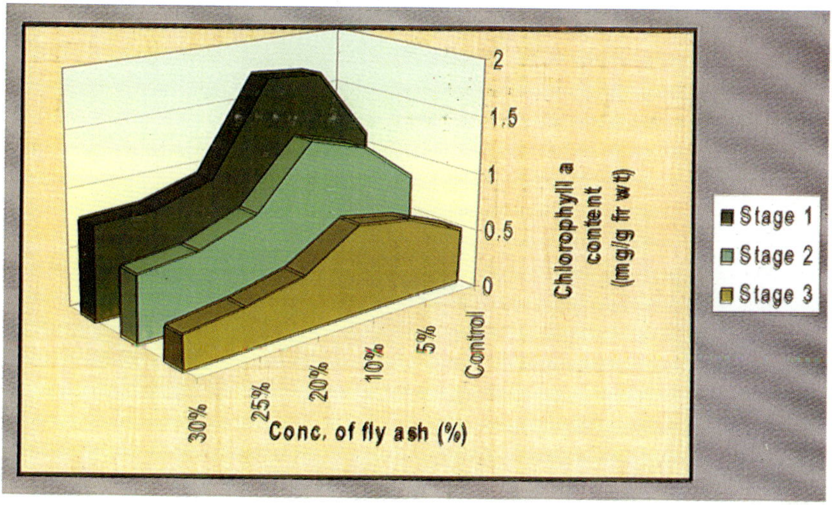

Figure 13: Comparison of chlorophyll a (mg gm^{-1} fr. wt.) of the leaves of *M. alba* during different stages of growth and at various fly ash concentrations in soil

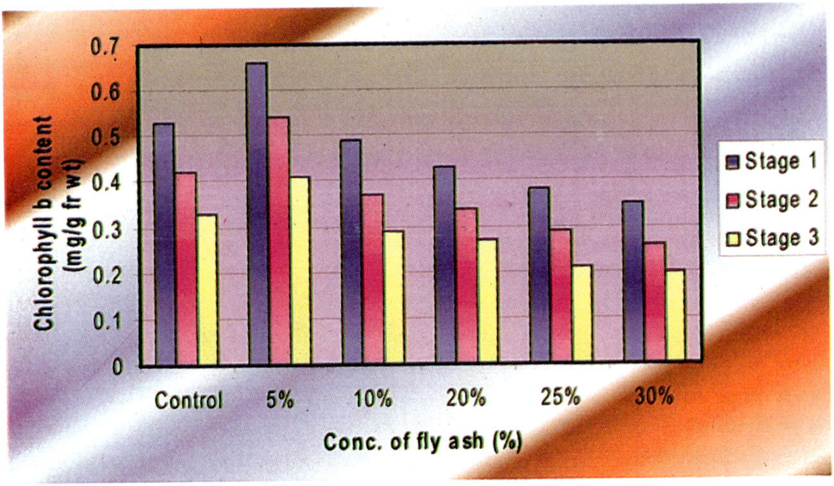

Figure 14: Comparison of chlorophyll b (mg gm^{-1} fr. wt.) of the leaves of *C. dichotoma* during different stages of growth and at various fly ash concentrations in soil

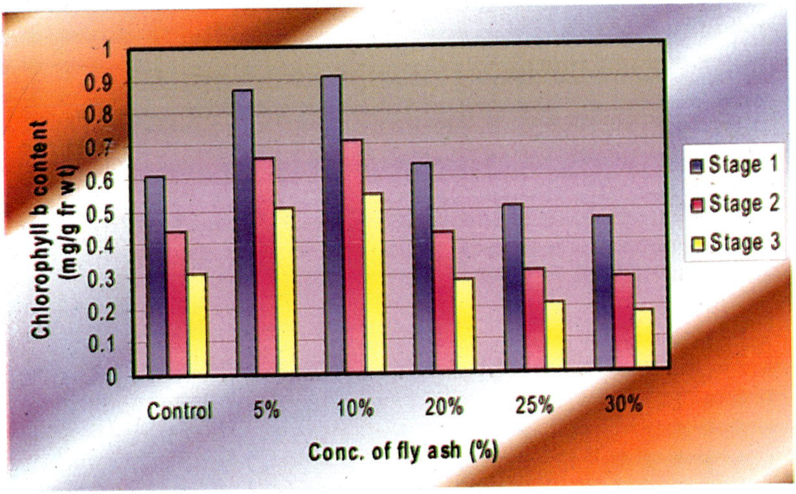

Figure 15: Comparison of chlorophyll b (mg gm^{-1} fr. wt.) of the leaves of *M. alba* during different stages of growth and at various fly ash concentrations in soil

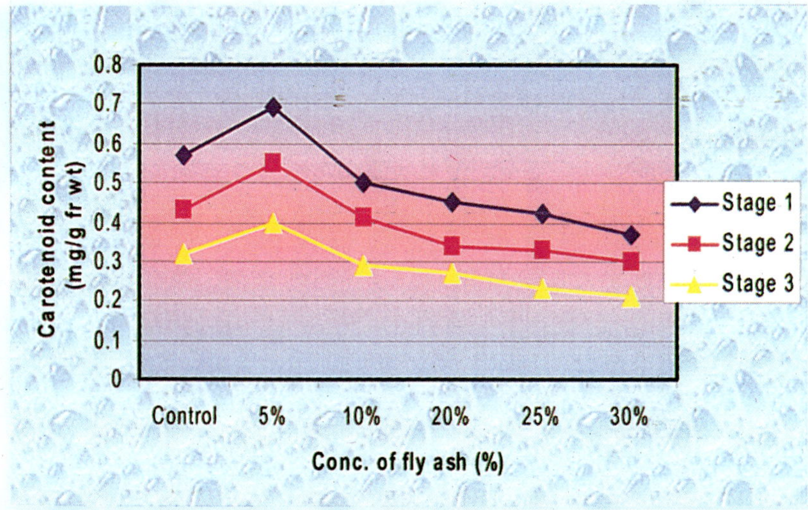

Figure 16: Comparison of carotenoids (mg gm^{-1} fr. wt.) of the leaves of *C. dichotoma* during different stages of growth and at various fly ash concentrations in soil

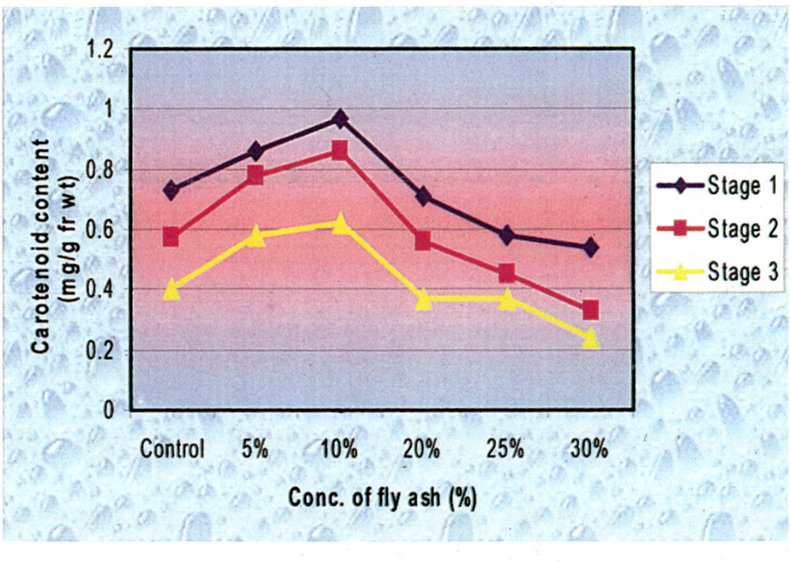

Figure 17: Comparison of carotenoids (mg gm^{-1} fr. wt.) of the leaves of *M. alba* during different stages of growth and at various fly ash concentrations in soil

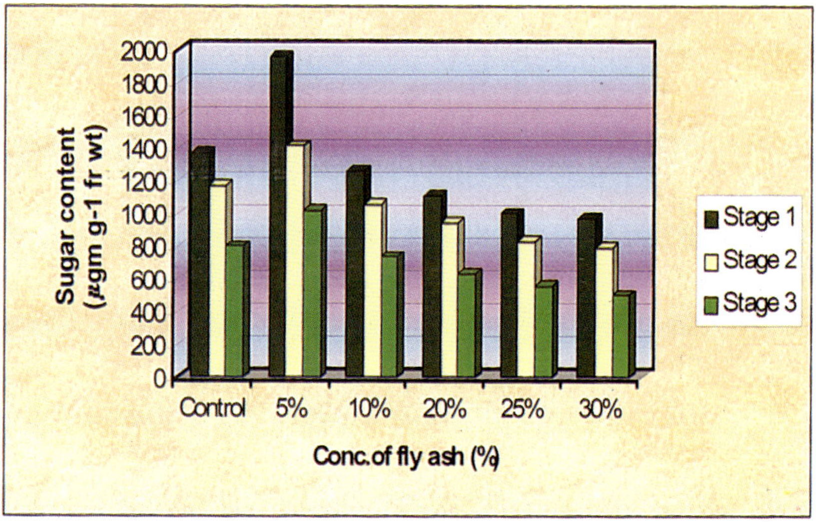

Figure 18: Comparison of sugar contents (μ gm g^{-1} fr. wt.) of the leaves of *C. dichotoma* during different stages of growth and at various fly ash concentrations in soil

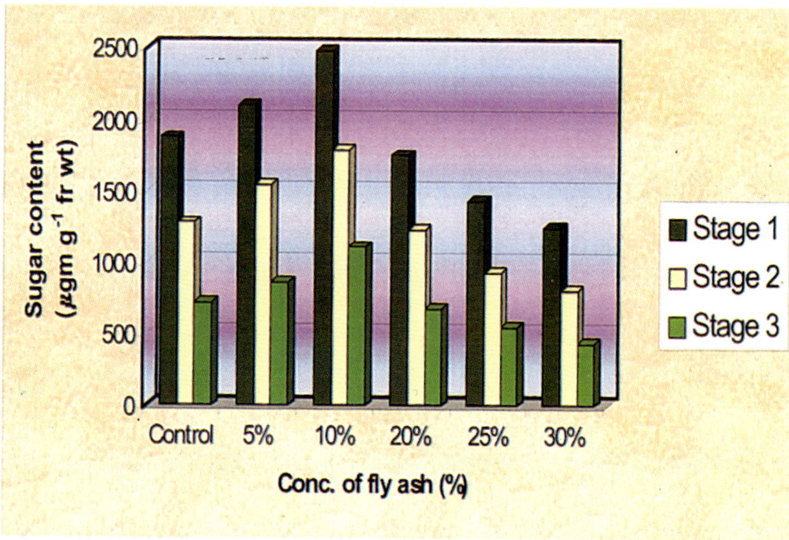

Figure 19: Comparison of sugar content (μ gm g^{-1} fr. wt.) of the leaves of *M. alba* during different stages of growth and at various fly ash concentrations in soil.

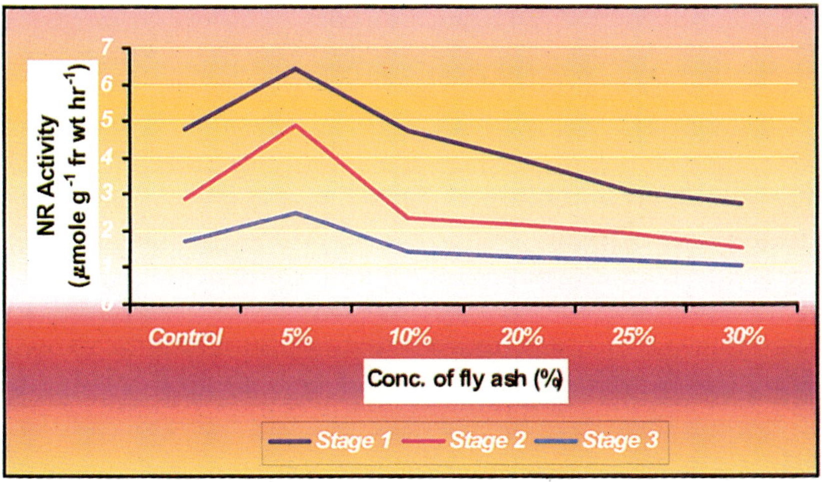

Figure 20: Comparison of NR activity (µ mole g^{-1} fr. wt. hr^{-1}) in the leaves of *C. dichotoma* during different stages of growth and at various fly ash concentrations in soil

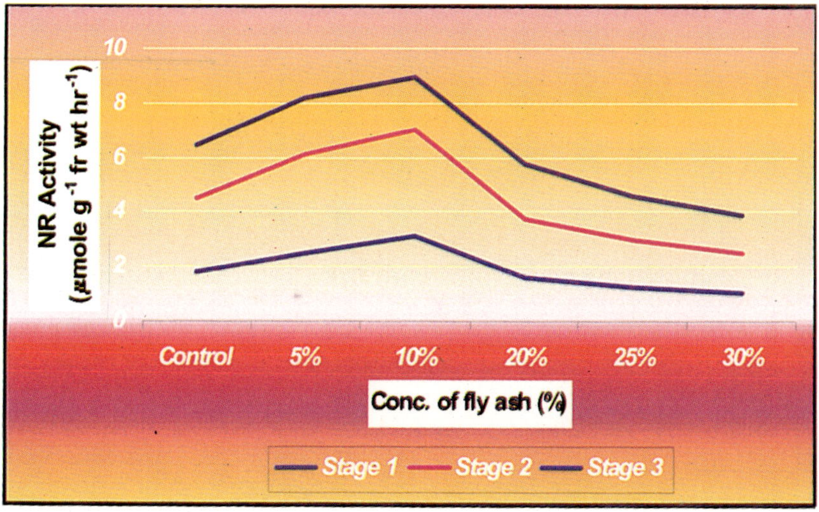

Figure 21: Comparison of NR activity (µ mole g^{1} fr. wt. hr^{-1}) in the leaves of *M. alba* during different stages of growth and at various fly ash concentrations in soil.

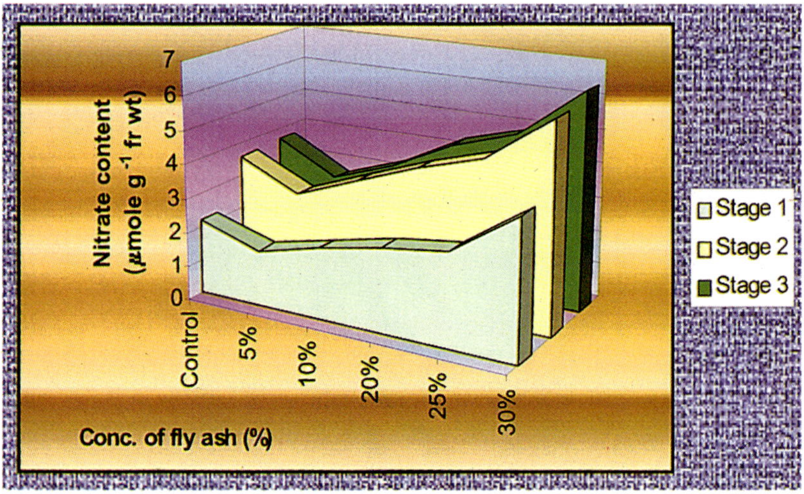

Figure 22: Comparison of nitrate contents (μ mole g[1] fr. wt.) in the leaves of *C. dichotoma* during different stages of growth and at various fly ash concentrations in soil.

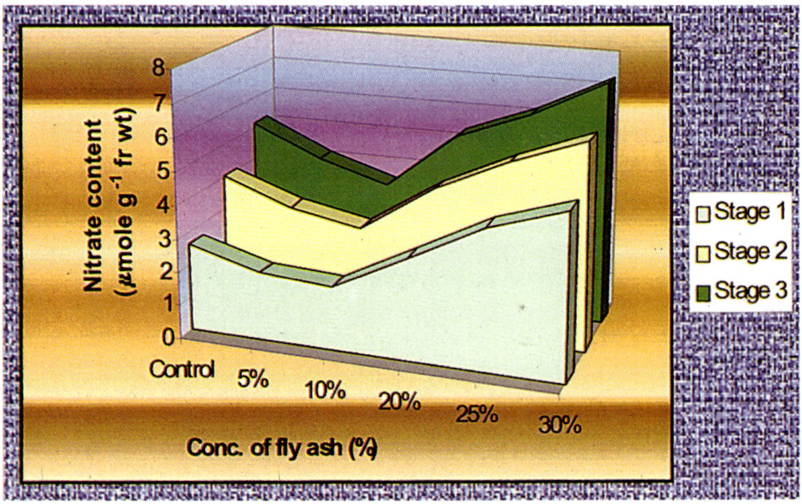

Figure 23: Comparison of nitrate content (μ mole g^{-1} fr.wt.) in the leaves of *M. alba* during different stages of growth and at various fly ash concentrations in soil.

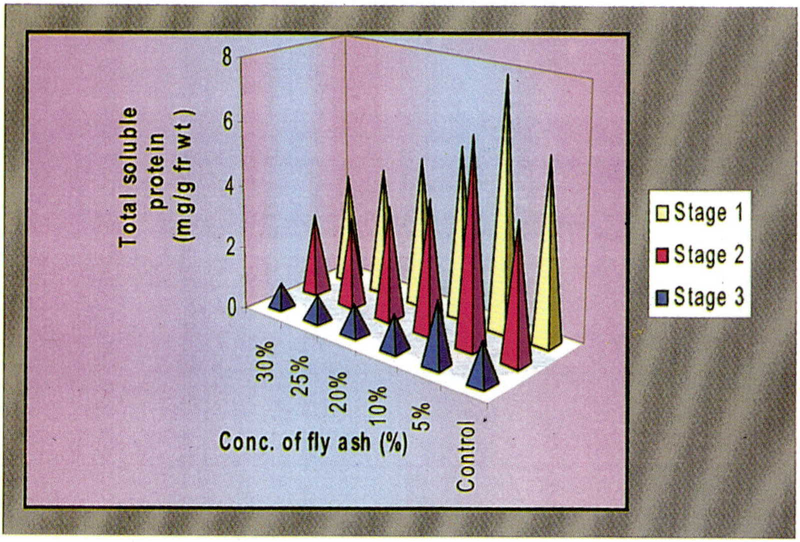

Figure 24: Comparison of protein content (mg g^{-1} fr.wt.) of the leaves of *C. dichotoma* during different stages of growth and at various fly ash concentrations in soil.

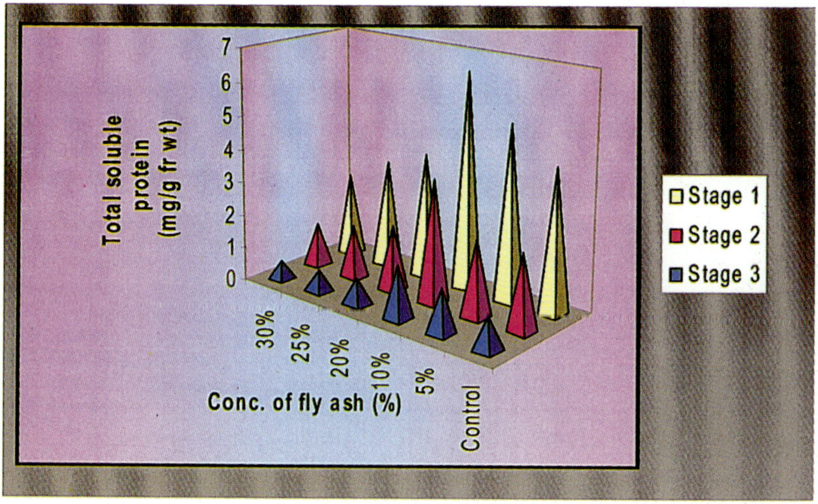

Figure 25: Comparison of protein content (mg g^{-1} fr.wt.) of the leaves of *M. alba* during different stages of growth and at various fly ash concentrations in soil.

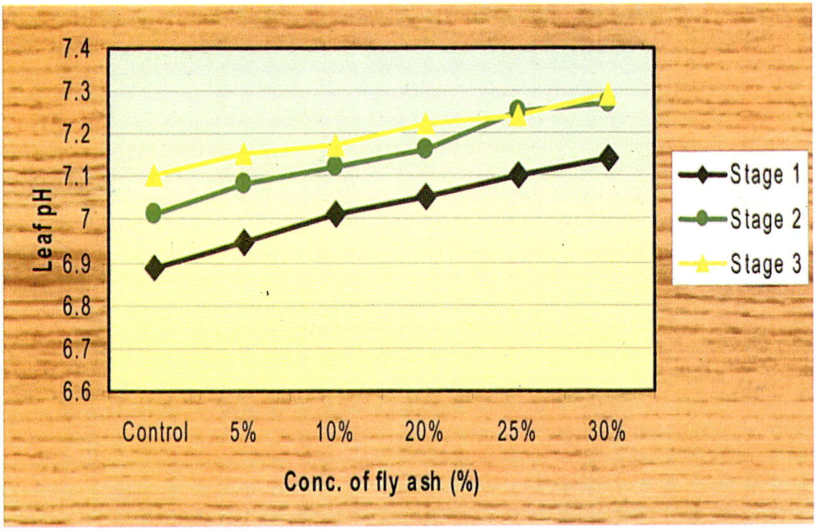

Figure 26: Comparison of leaf pH of *C. dichotoma* during different stages of growth and at various fly ash concentration in soil.

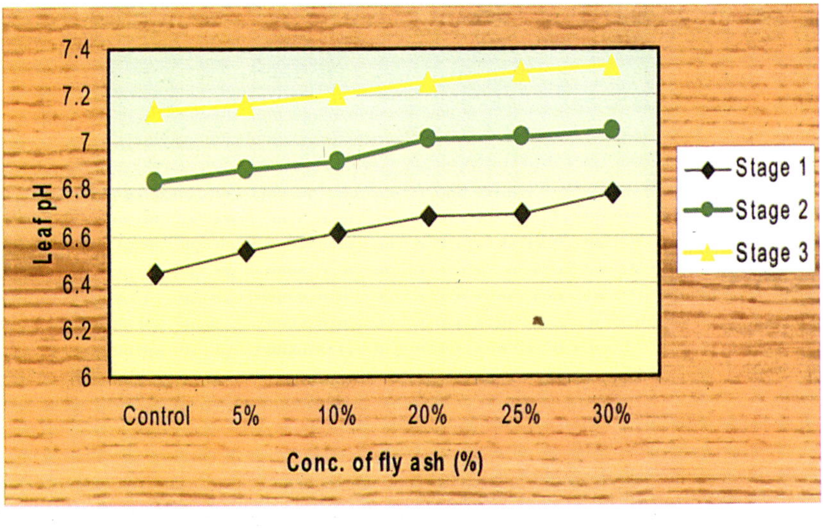

Figure 27: Comparison of leaf pH of *M. alba* during different stages of growth and at various fly ash concentrations in soil

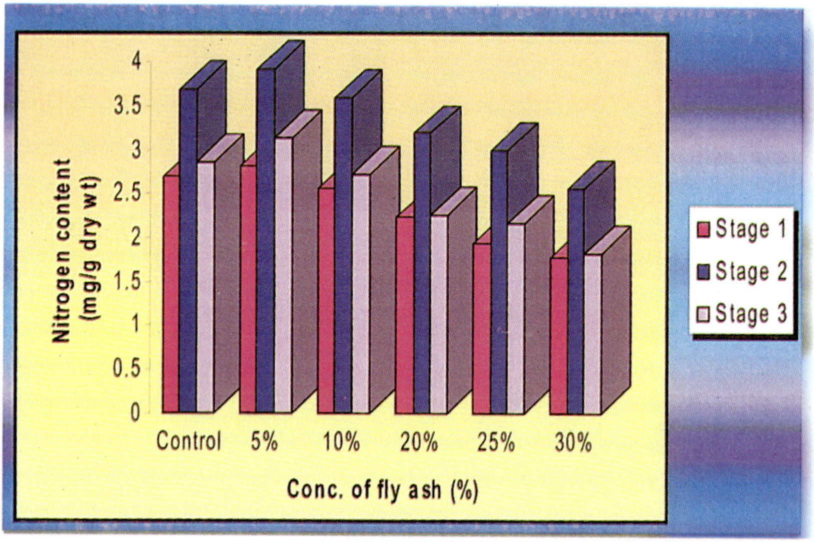

Figure 28: Comparison of nitrogen content (mg g^{-1} dry wt.) of the leaves of *C. dichotoma* during different stages of growth and at various fly ash concentrations in soil.

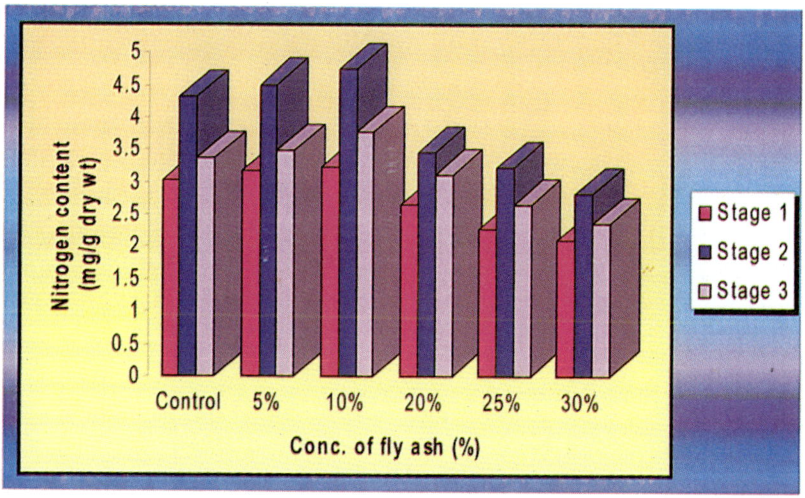

Figure 29: Comparison of nitrogen content (mg g^{-1} dry wt.) of the leaves of *M. alba* during different stages of growth and at various fly ash concentrations in soil.

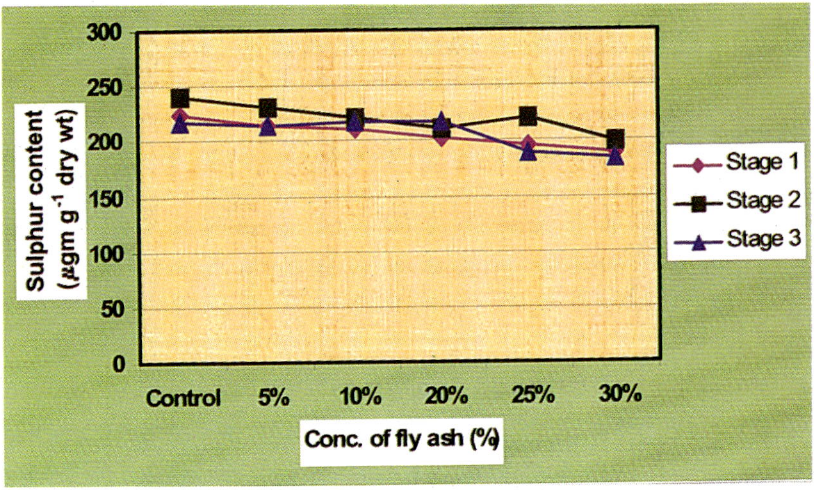

Figure 30: Comparison of Sulphur Content (μ gm g^{-1} dry wt.) of the leaves of *C. dichotoma* during different stages of Growth and at Various Fly Ash Concentrations in Soil.

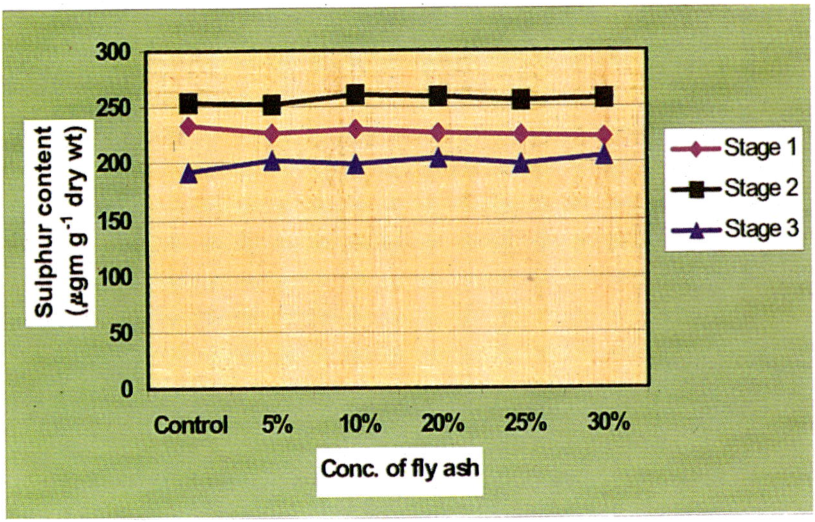

Figure 31: Comparison of Sulphur content (μ gm g^{-1} dry wt.) of the leaves of *M. alba* during different stages of grwoth and at various fly ash concentrations in soil.

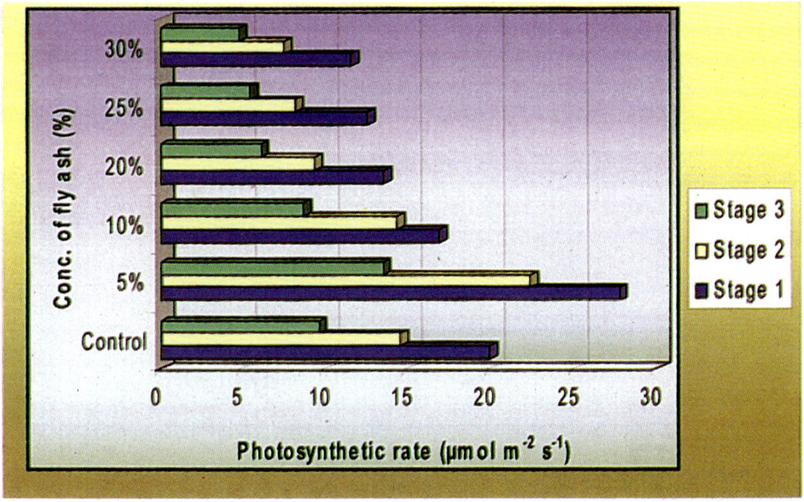

Figure 32: Comparison of photosynthetic rate (µ mole CO_2 m^{-2} s^{-1}) of the leaves of *C. dichotoma* during different stages of growth and at various fly ash concentrations in soil.

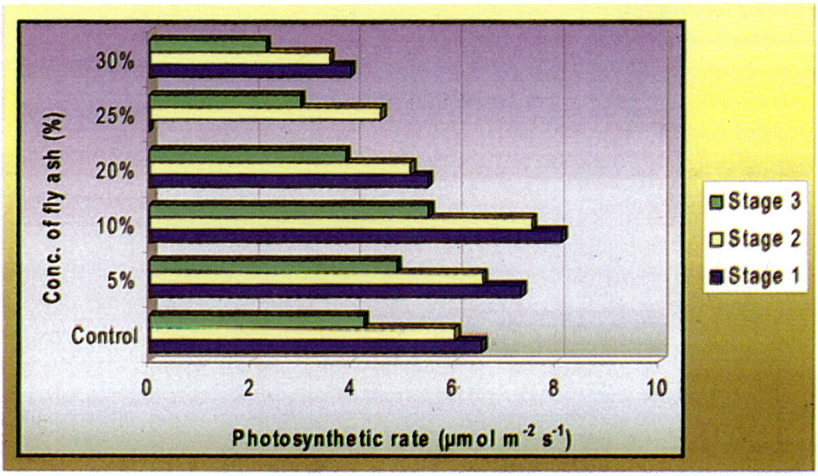

Figure 33: Comparison of photosynthetic rate (µ mole CO_2 m^{-2} s^{-1}) of the leaves of *M. alba* during different stages of growth and at various fly ash concentrations in soil.

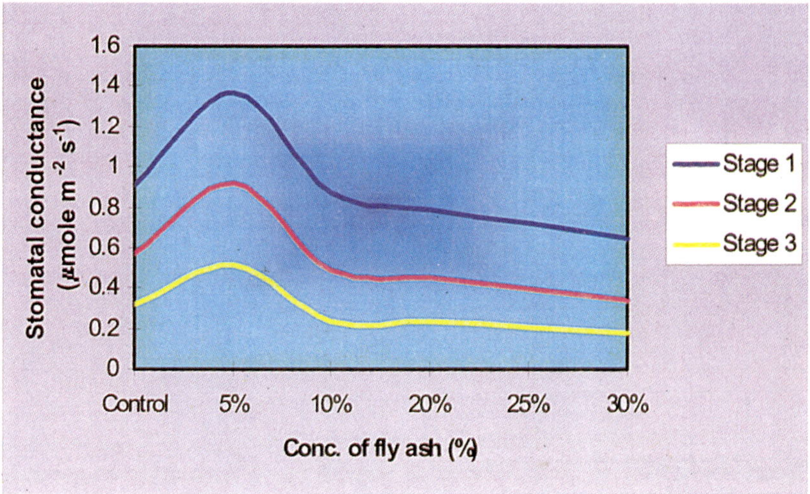

Figure 34: Comparison of stomatal conductance (µ mole m^{-2} s^{-1}) of the leaves of *C. dichotoma* during different stages of growth and at various fly ash concentrations in soil.

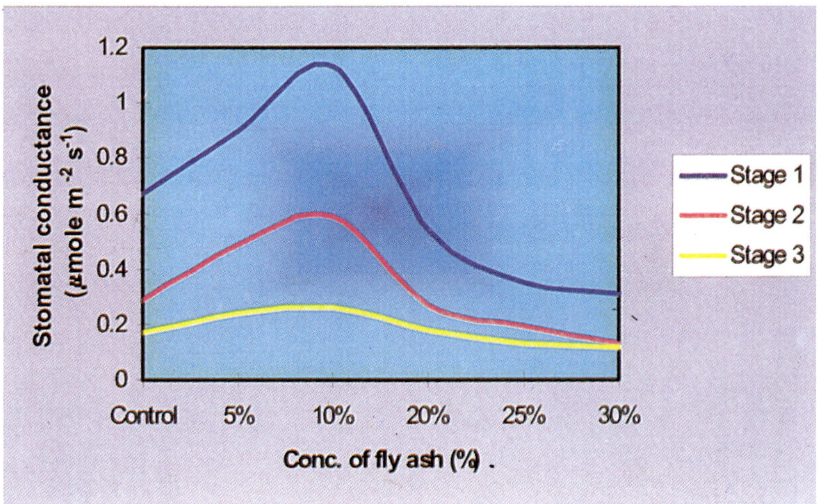

Figure 35: Comparison of stomatal conductance (µ mole m^{-2} s^{-1}) of the leaves of *M. alba* during different stages of growth and at various fly ash concentrations in soil.

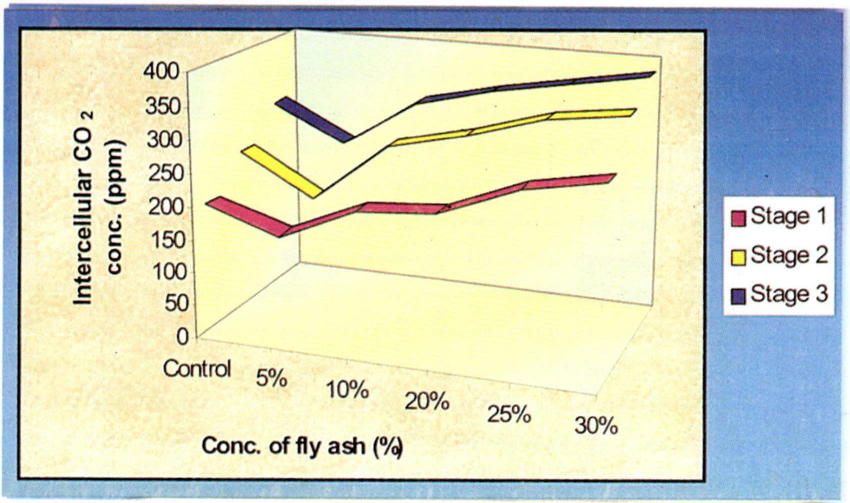

Figure 36: Comparison of intercellular CO_2 conc. (ppm) of the leaves of *C. dichotoma* during different stages of grwoth and at various fly ash concentrations in soil.

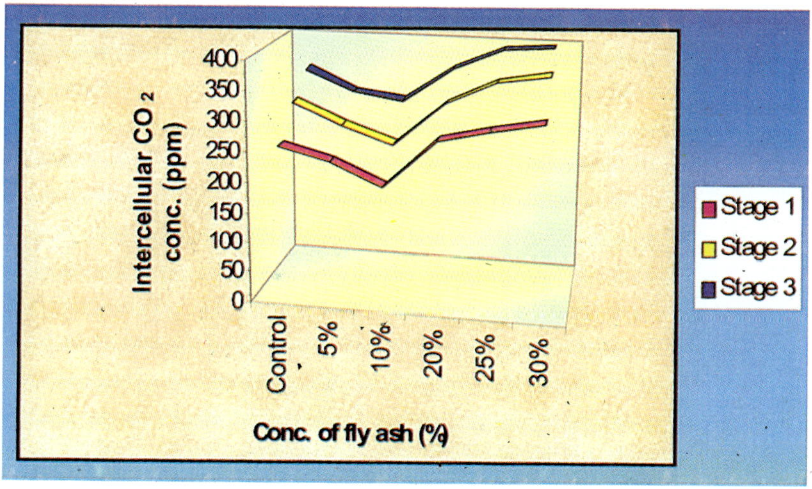

Figure 37: Comparison of intercellular CO_2 conc. (ppm) of the leaves of *M. alba* during different stages of grwoth and at various fly ash concentrations in soil.

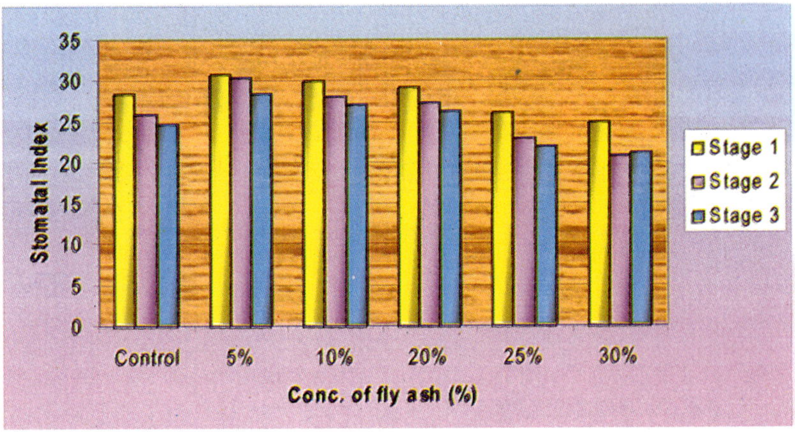

Figure 38: Comparison of stomatal index of abaxial epidermis of the leaves of *C. dichotoma* during different stages of grwoth and at various fly ash concentrations in soil.

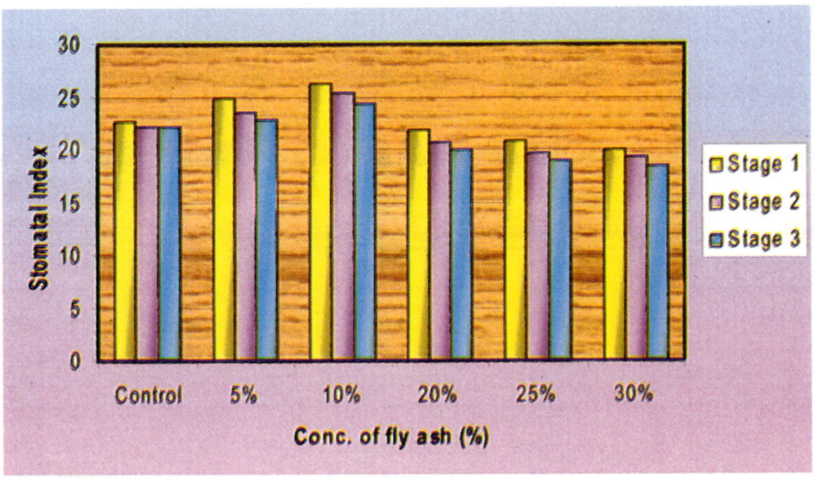

Figure 39: Comparison of stomatal index of abaxial epidermis of the leaves of *M. alba* during different stages of grwoth and at various fly ash concentrations in soil.

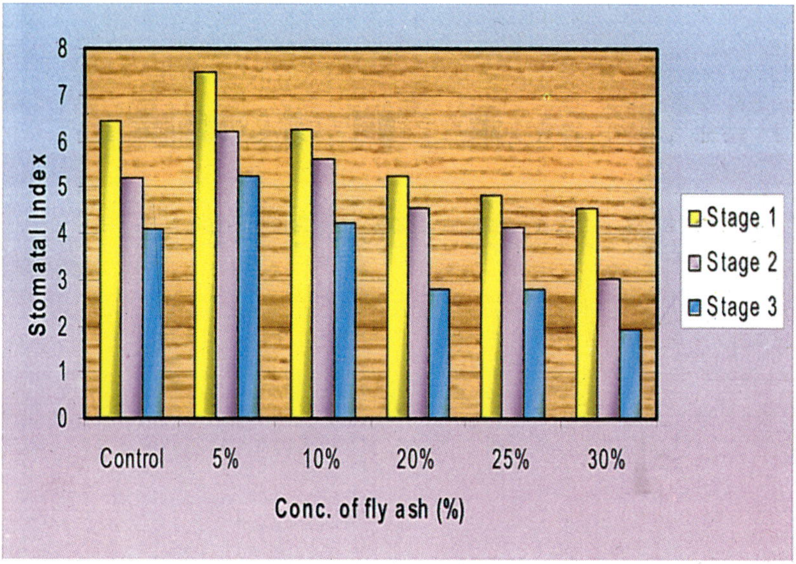

Figure 40: Comparison of stomatal index of adaxial epidermis of the leaves of *C. dichotoma* during different stages of grwoth and at various fly ash concentrations in soil.

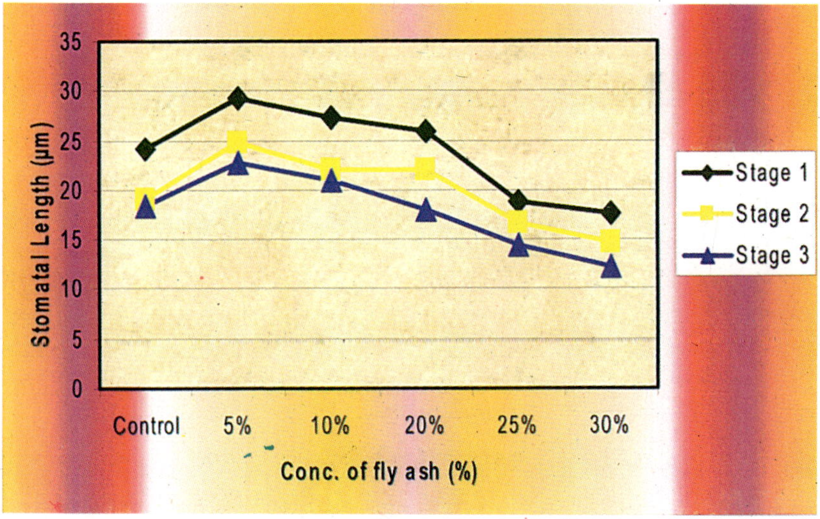

Figure 41: Comparison of stomatal length (µ m) of abaxial epidermis of the leaves of *C. dichotoma* during different stages of growth and at various fly ash concentrations in soil.

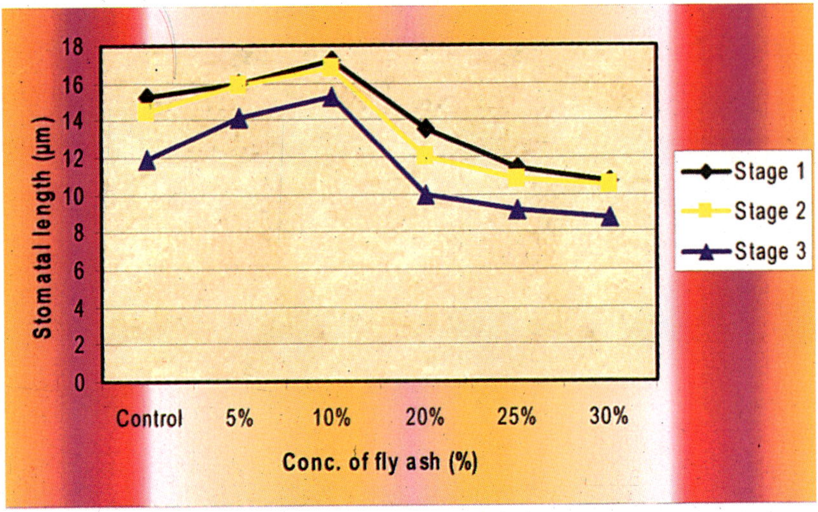

Figure 42: Comparison of stomatal length (μ m) of abaxial epidermis of the leaves of *M. alba* during different stages of growth and at various fly ash concentrations in soil.

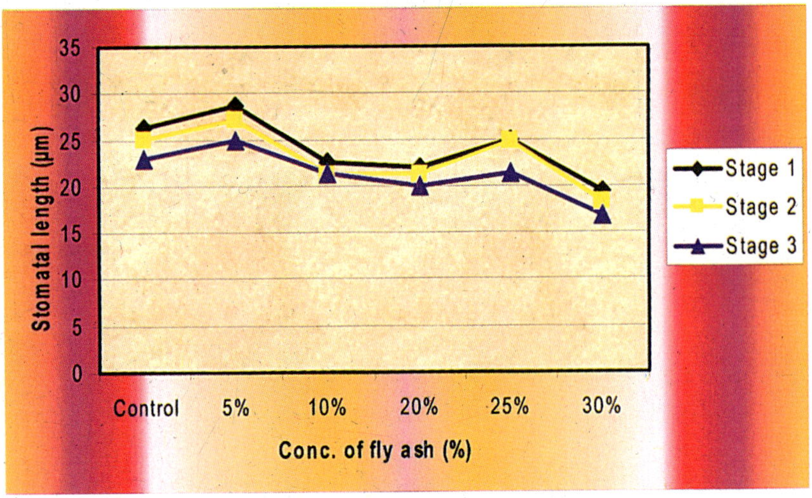

Figure 43: Comparison of stomatal length (μ m) of adaxial epidermis of the leaves of *C. dichotoma* during different stages of growth and at various fly ash concentrations in soil.

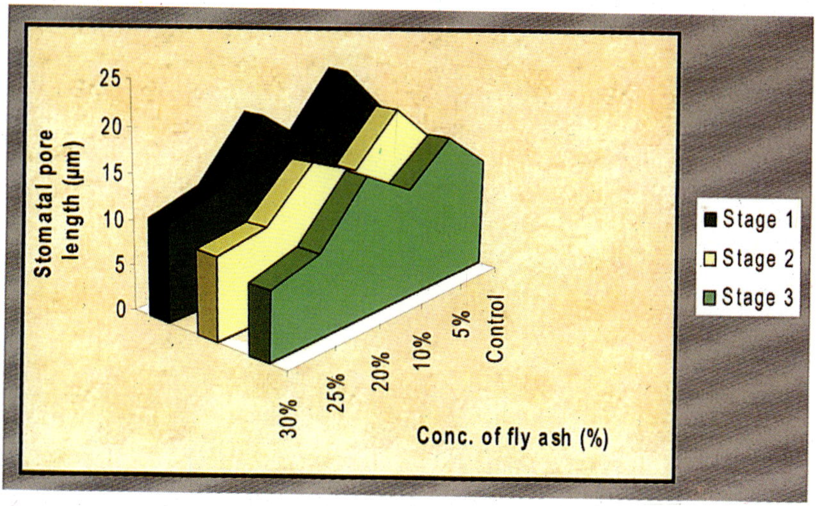

Figure 44: Comparison of stomatal pore length (µ m) of abaxial epidermis of the leaves of *C. dichotoma* during different stages of growth and at various fly ash concentrations in soil.

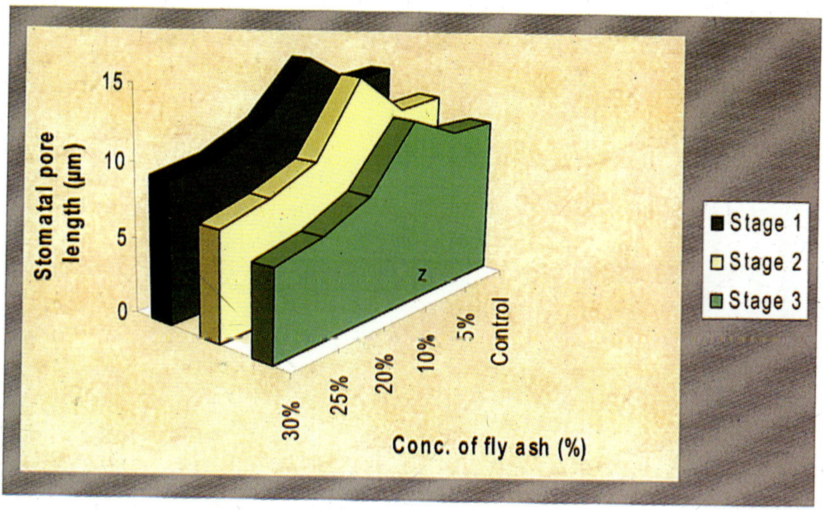

Figure 45: Comparison of stomatal pore length (µ m) of abaxial epidermis of the leaves of *M. alba* during different stages of growth and at various fly ash concentrations in soil.

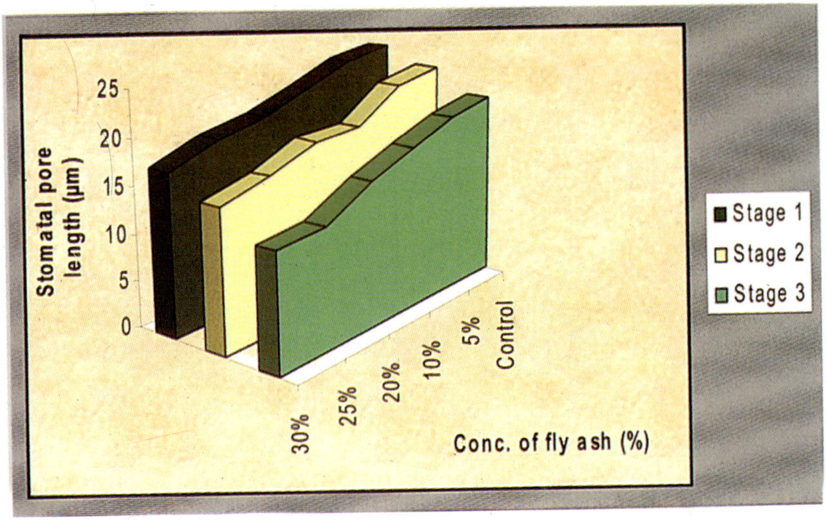

Figure 46: Comparison of stomatal pore length (μ m) of adaxial epidermis of the leaves of *C. dichotoma* during different stages of growth and at various fly ash concentrations in soil.

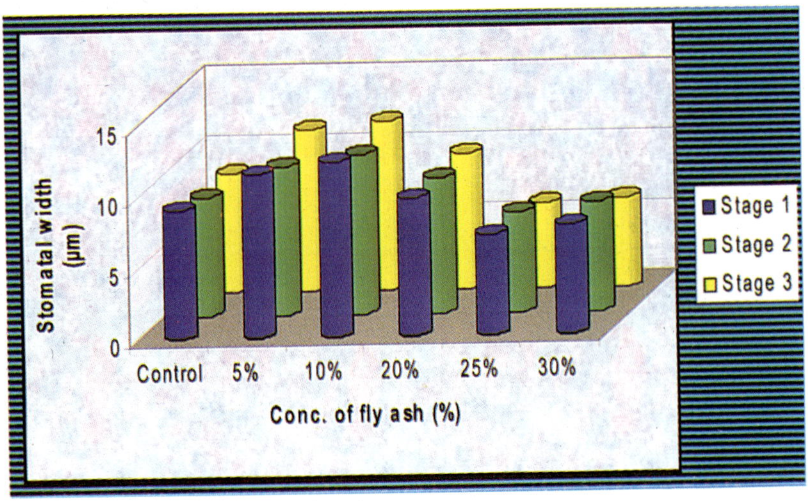

Figure. 47: Comparison of stomatal width (μ m) of abaxial epidermis of the leaves of *C. dichotoma* during different stages of growth and at various fly ash concentrations in soil.

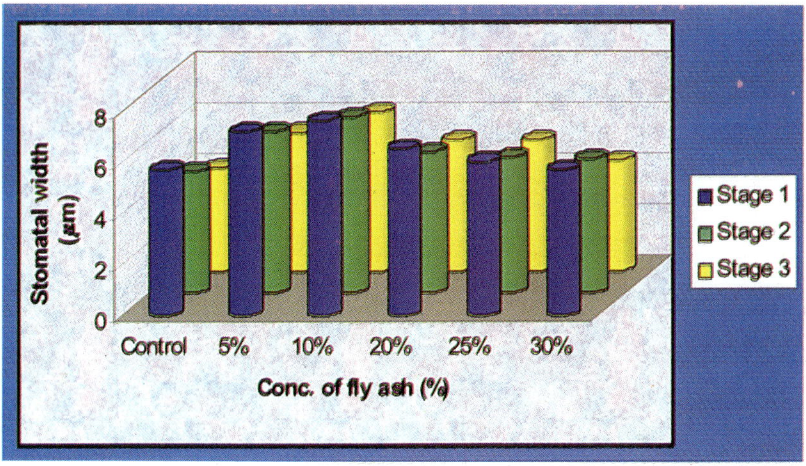

Figure 48: Comparison of stomatal width (μ m) of abaxial epidermis of the leaves of *M. alba* during different stages of growth and at various fly ash concentrations in soil.

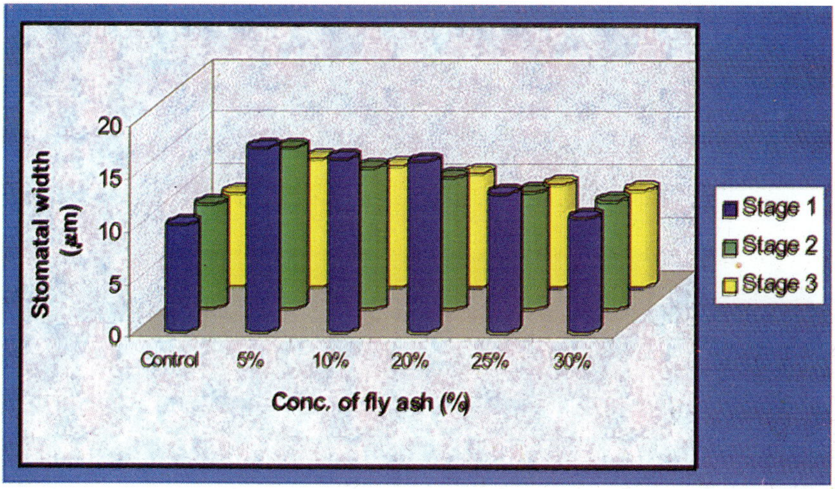

Figure 49: Comparison of stomatal width (μ m) of adaxial epidermis of the leaves of *C. dichotoma* during different stages of growth and at various fly ash concentrations in soil.

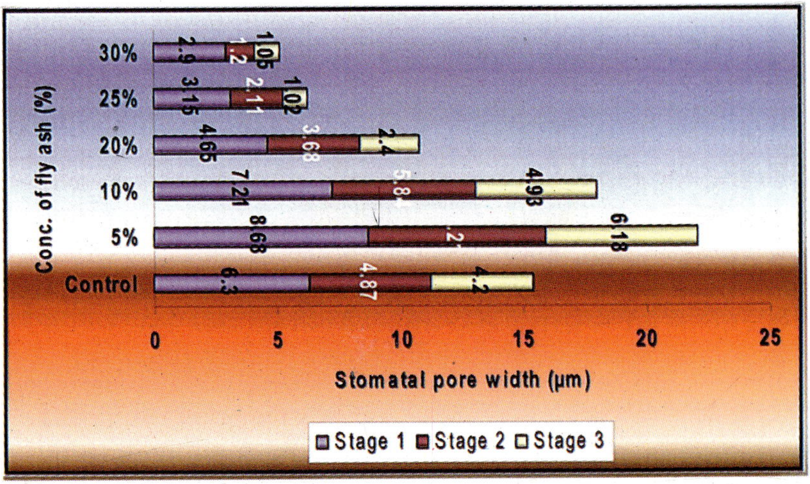

Figure 50: Comparison of stomatal pore width (μ m) of abaxial epidermis of the leaves of *C. dichotoma* during different stages of growth and at various fly ash concentrations in soil.

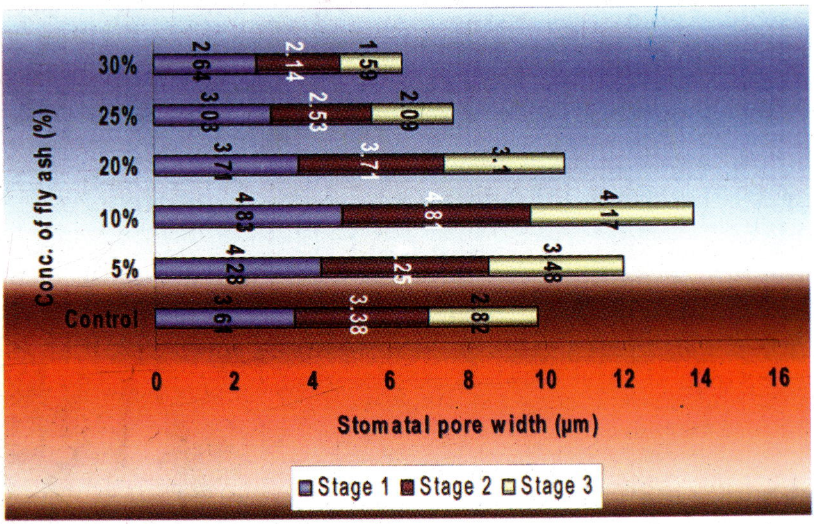

Figure 51: Comparison of stomatal pore width (μ m) of abaxial epidermis of the leaves of *M. alba* during different stages of growth and at various fly ash concentrations in soil.

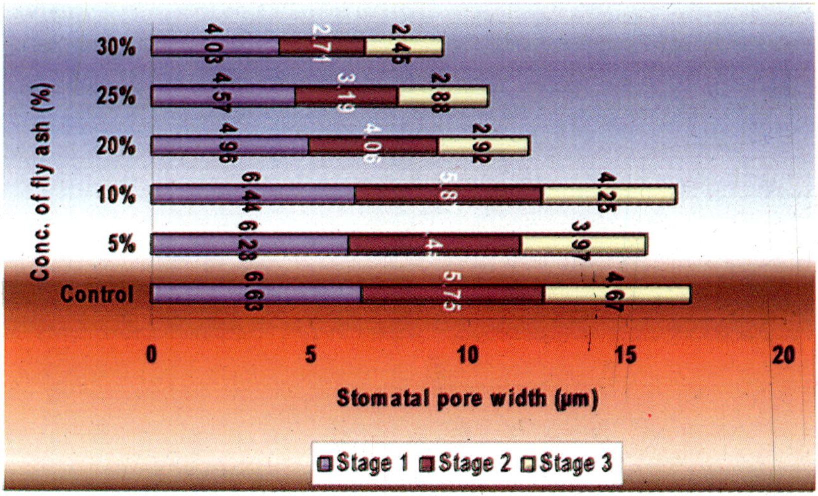

Figure 52: Comparison of stomatal pore width (μ m) of adaxial epidermis of the leaves of *C. dichotoma* during different stages of growth and at various fly ash concentrations in soil.

PLATE 1

Badarpur thermal power station (a), the source of generation of a variety of pollutants including fly ash. The ash pond of lattice design where the waste fly ash is dumped from the power station (b).

PLATE 2

The photographs show control leaves (a) and leaves of 5% fly ash treated soil (b) of *C. dichotoma*. Maximum leaf size and flourished growth was reported at 5% fly ash treatment of soil as against control leaves.

PLATE 3

The size of *C. dichotoma* leaves decreased as the concentration of fly ash in soil increased starting from 10% (a), the maximum reduction however, was reported at 30% fly ash dose. The interveinal chlorotic patches started showing up at 20% fly ash levels (b) that increased in frequency with visible leaf burns at 25% and 30% leaves (c and d).

PLATE 4

The leaf area of *M. alba* increased with increasing concentrations up to 5% and 10% treatment of fly ash as against control leaves (a). The significant increase in size of leaves was however reported at 5% (b) and 10% (c) fly ash levels.

PLATE 5

With the further increasing concentration of (20%, 25% and 30%) the size of *M. alba* leaves decreased. Marginal leaf chlorosis and leaf burn was evident with 20% treatment of fly ash (a). The chlorosis spread to entire leaf and the tip burn extended to peripheral necrosis at 25% fly ash level (b). The chlorotic patches started turning up to be necrotic at 30% fly ash dose (c).

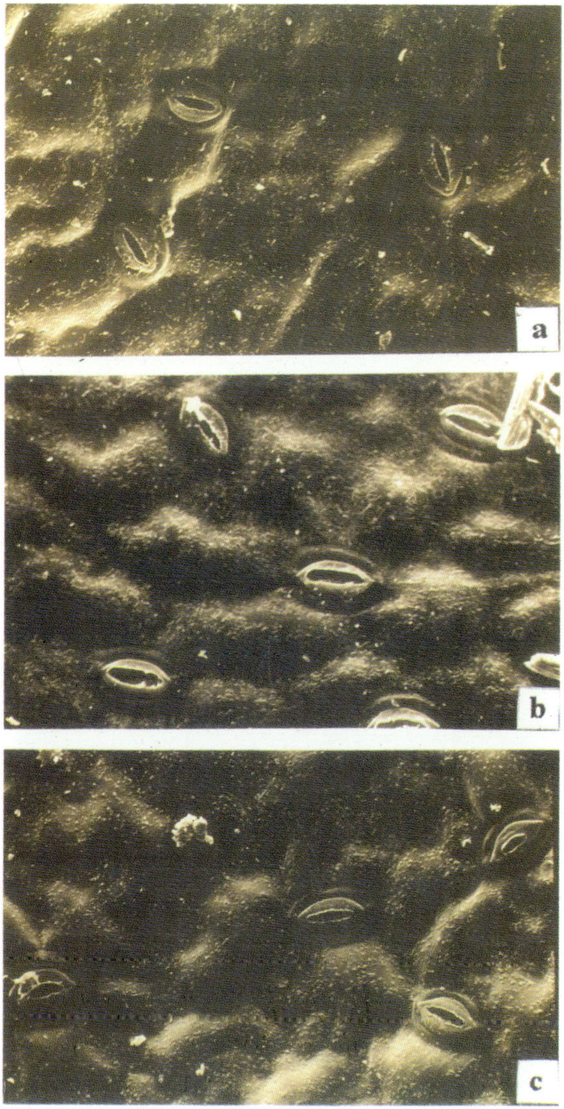

PLATE 6

Electro micrograph of the lower epidermis of *C. dichotoma* leaves showing the size distribution of stomata. The maximum stomatal density is observed at 5% fly ash the (b). The size of stomatal pore is also greatest with 5% fly ash level (b) while at 10% level (c) and control leaves (a) the stomatal size and density is low. Magnification 650 X

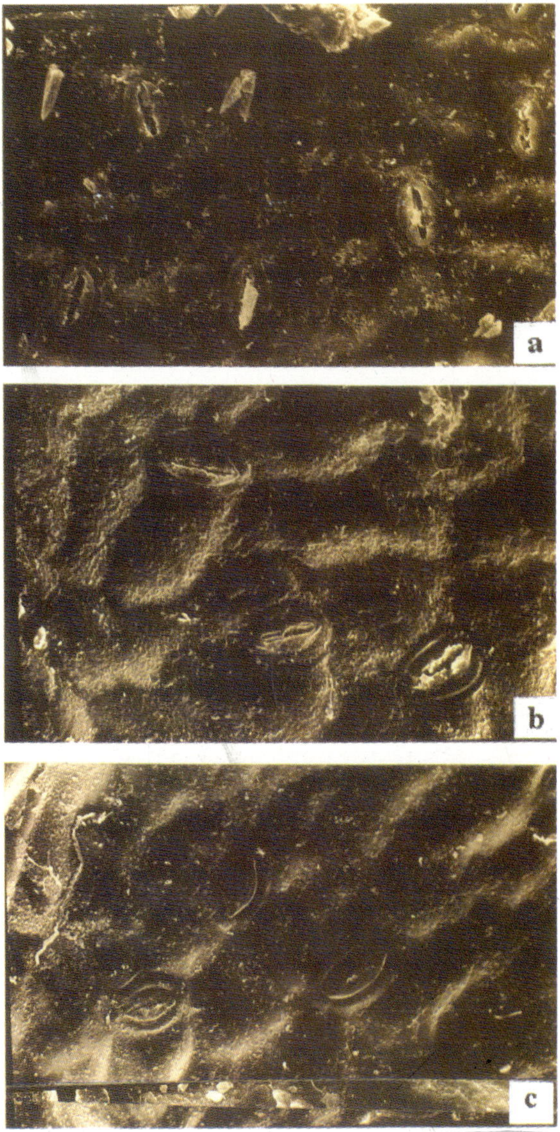

Electro micrographs of the lower epidermis of *C. dichotoma* leaves showing the reduced aperture size of 25% and 30% fly ash treated leaves (b and c). The stomatal aperture is open at 20% level (a) though there are a few blockages but it is completely closed at 30% level (c).
Magnification : 650X.

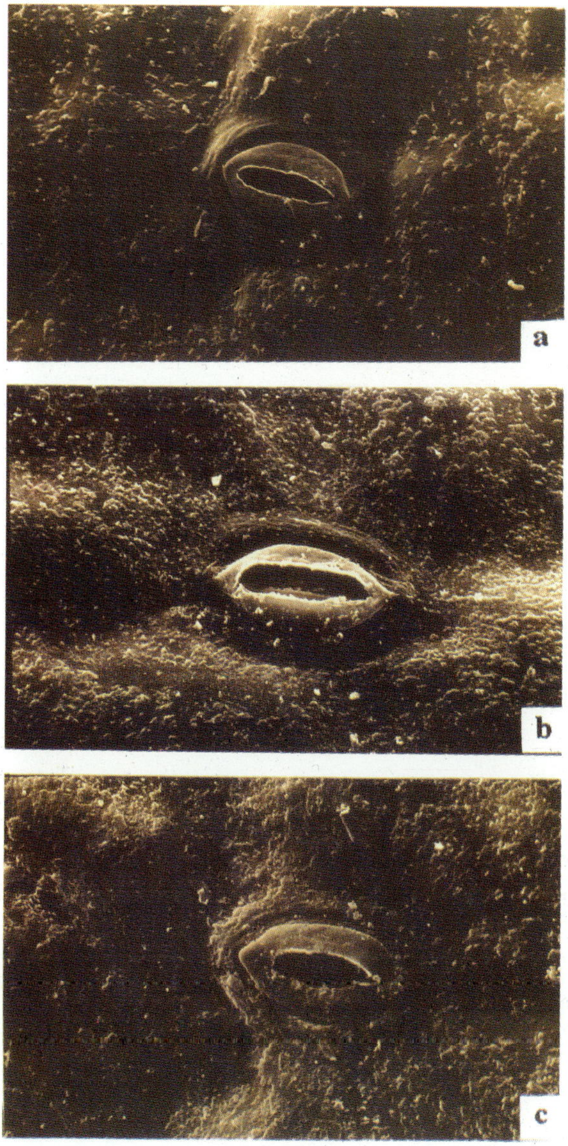

PLATE 8

Electro micrographs of the lower epidermis of C. *dichotoma* leaves showing the increase in stomatal aperture length and width at 5% and 10% level of fly ash as against control leaves (a). The stomatal length and stomatal pore length was recorded maximum with 5% fly ash dose (b) however a marked increase in stomatal width and aperture width occurred at 10% fly ash level (c). Magnification 1.5 KX

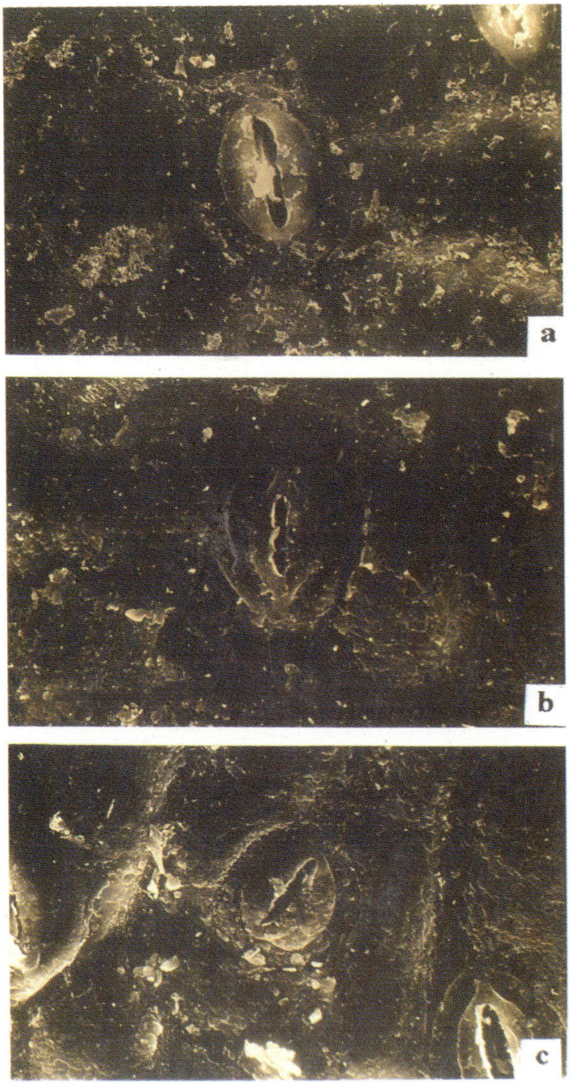

PLATE 9

Electro micrographs of the lower epidermis of C. *dichotoma* leaves showing the variation in size of stomata and stoma at 20%, 25% and 30% fly ash levels. The decline in size of stomata and stoma was evident at 20% treatment of fly ash (a) than of control leaves The reduction in stomatal length and aperture length was maximum at 30% fly ash dose (c) while the stomatal width and stomatal pore width decreased maximally at 25% fly ash level (b).

Magnification 1.5 KX

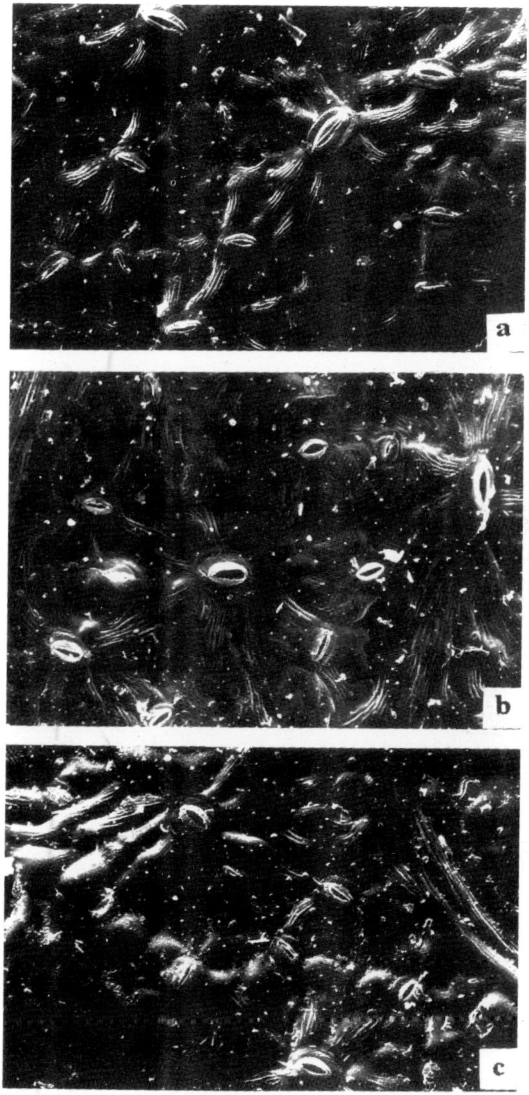

PLATE 10

Electro micrographs of the lower epidermis of *M. alba* showing the stomatal distribution and stomatal size at control, 5% and 10% fly ash levels. Control leaves (a) were found to bear less stomata than at 5% (b) and 10% (c) fly ash levels. The maximum increase was recorded however with 10% level of fly ash (b).

Magnification: 650 X

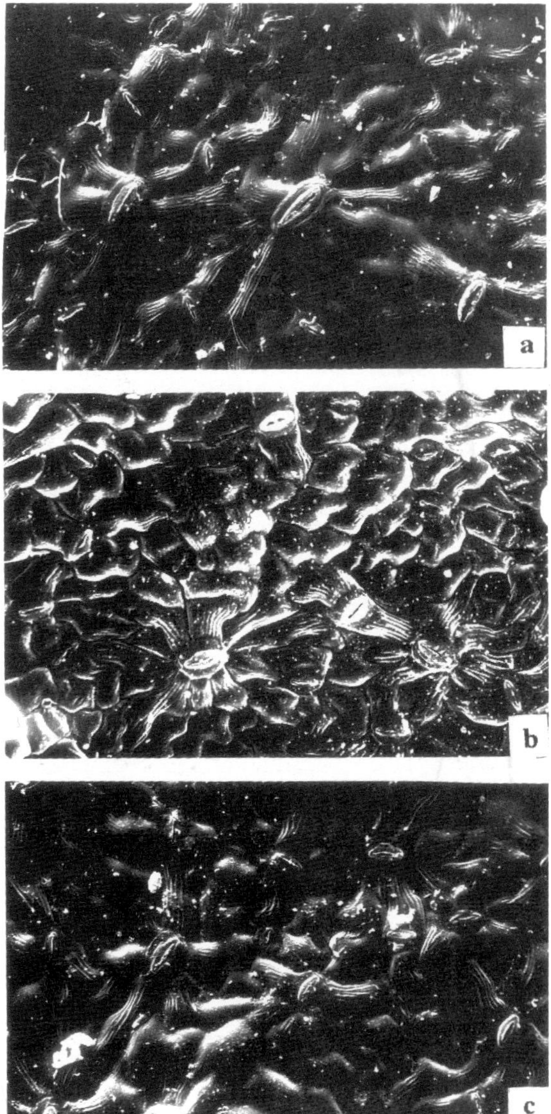

PLATE 11

Electro micrographs of the lower epidermis of *M. alba* showing a decrease in stoma index at 20%, 25% and 30% doses of fly ash. The stomatal index declined at 20% (a) and 25% (b) fly ash level, however the maximum reduction in stomatal index was observed with 30% treatment of fly ash(c).
Magnification: 650 X

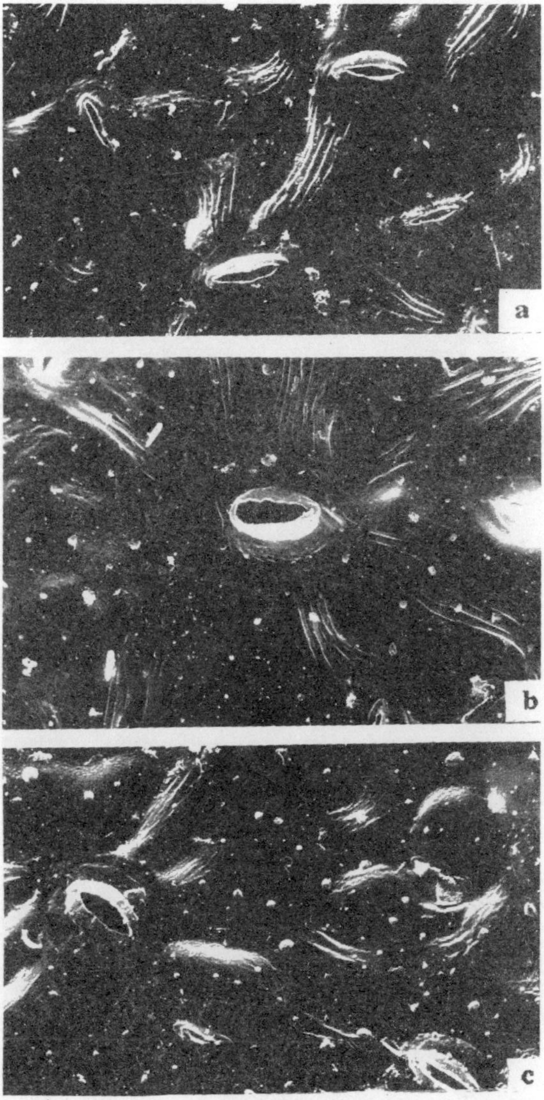

PLATE 12

Electro micrographs of the lower epidermis of *M. alba* leaves showing the size of stomata and stomatal aperture. The stomatal size and stomatal pore size increased up to 10% fly ash levels. The size of stomata and stoma for control leaves (a) was less than with 5% (b) and 10% (c) treatments of fly ash, the maximum being at 10%.

Magnification: 1.5 KX

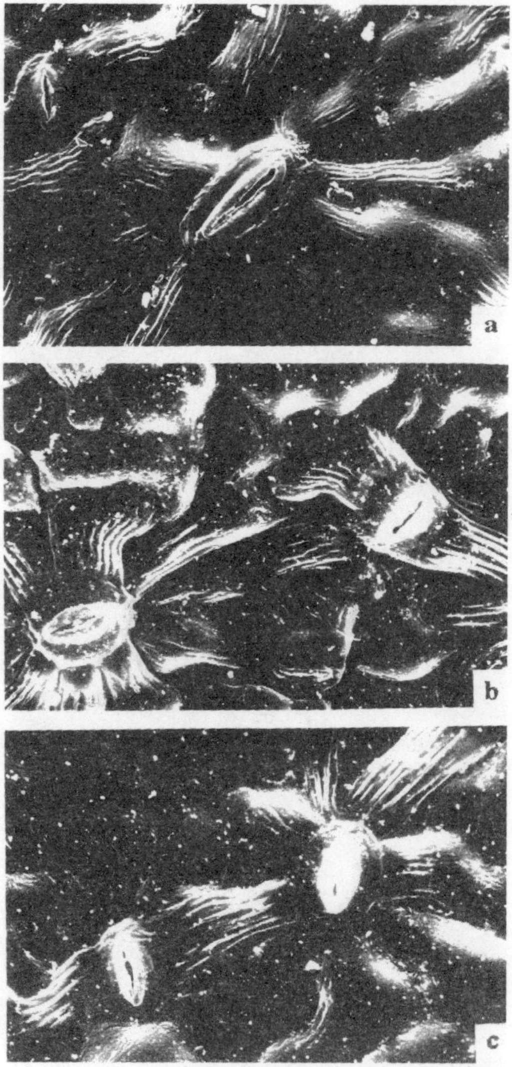

PLATE 13

Electro micrographs of the lower epidermis of *M. alba* leaves showing the size of stomata and stomatal aperture with 20% (a), 25% (b) and 30% (c) treatments of fly ash The stomatal length and aperture length decreased with the increasing concentrations of fly ash, the maximum being at 30% fly ash dose. The stomatal width and width of stomatal aperture however increased consistently with increasing fly ash concentrations.

Magnification: 1.5KX

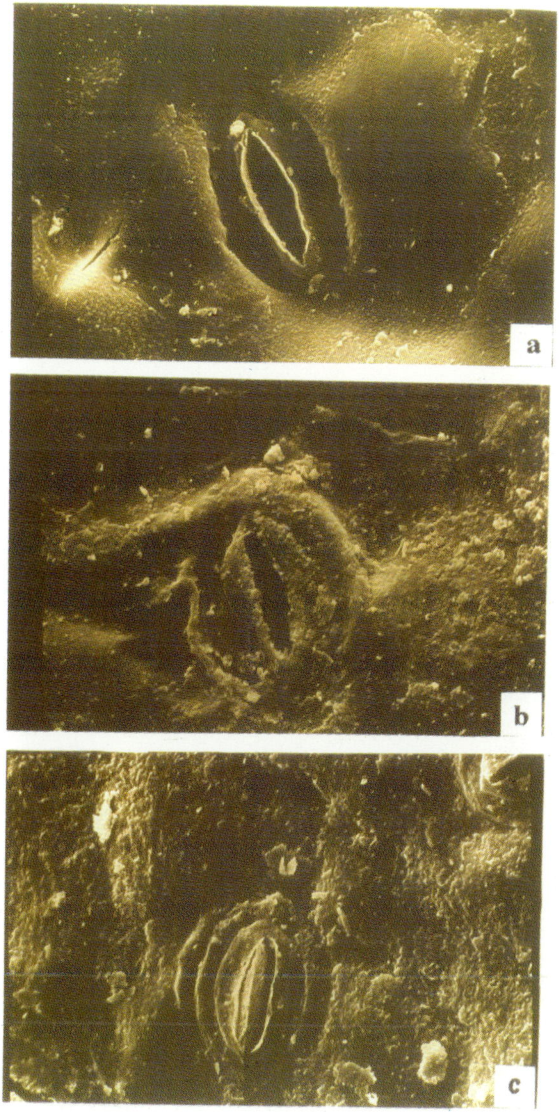

PLATE 14

Electro micrograph of the upper epidermis of *C. dichotoma* leaves depicting stomatal morphology and density. Stomatal length and stomatal width of control plants (a) was less than 5% fly ash treated plant (b). The stomatal pore length and pore width however decreased with 5% (b) and 10% (c) treated fly ash plants.
Magnification: 1.5KX

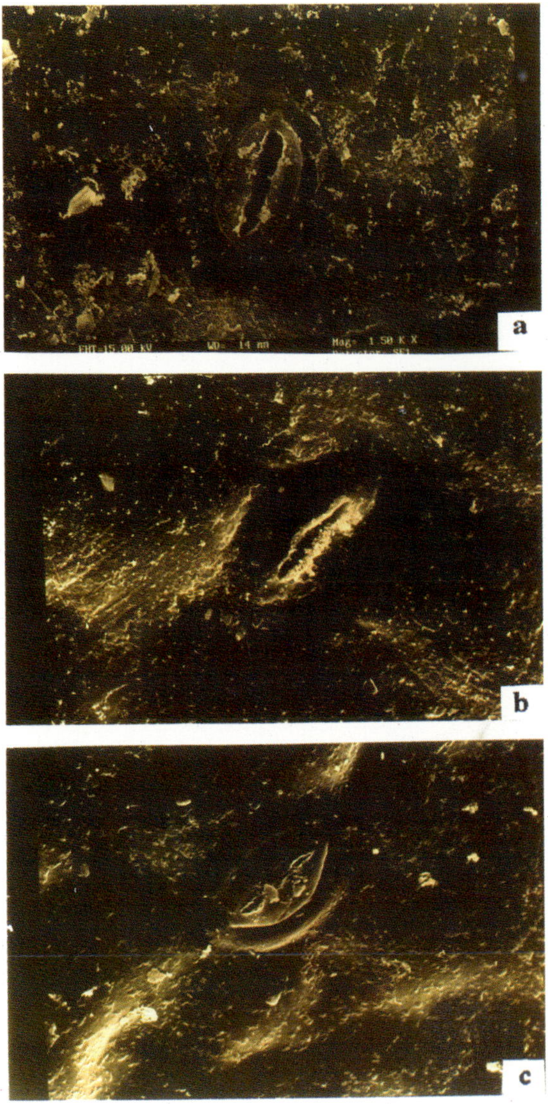

PLATE 15

Electro micrographs the upper epidermis of *C. dichotoma* leaves depicting stomatal morphology and density. The decline in stomatal length, stomatal pore length and stomatal pore width was evident with 20% (a), 25% (b) and 30% (c) treatments of fly ash in soil however the stomatal width increased as the concentration of fly ash in soil increased.

Magnification: 1.5KX

5

Discussion of the Results Obtained

Evaluation of heavy metals from fly ash in soils and plants is of great environmental importance due to a large amount of fly ash disposed and its effects on humans and other biota. All plant species under environmental conditions limit their performance (Osmand *et al.*, 1987; Nilsen and Orcutt, 1996). Various plant species including trees, vegetable crops, grasses and weeds are known to accumulate a wide range of heavy metals (Kabata - Pendias and Pendias, 1992). On one hand there are public health implications if plant foods accumulate high concentrations of heavy metals while on the other hand plants are being examined as an extractive tool for reclamation of metal contaminated fly ash dumps (Mishra *et al.*, 1995; Srivastava *et al.*, 1995; Williams *et al.*, 1995; Lal, 1996; Petrikova *et al.*, 1996).

The conventional method of disposal of fly ash is to transfer it as slurry to ash ponds and thereby biologically reclaiming the ash pond through phytoremediation by identifying plants that tolerate and absorb high levels of metals (Zhu *et al.*, 1999; Pichtel *et al.*, 2000). The phytoremediation would help in cleaning of contaminated soils by restoring its fertility and thereby its possible reuse (Robinson *et al.*, 1997; Robinson *et al.*, 1998; Frey *et al.*, 2000; Perronnet *et al.*, 2000).

Amendment of soils through addition of fly ash is yet another method for the beneficial use of this waste product. Fly ash can supply essential elements to nutrient deficient soils (Francis *et al.*, 1985; Nass *et al.*, 1993). Amelioration of agriculture soil with fly ash can modify soil properties thus improving the soil fertility and crop yield (Cervelli *et al.*, 1986; Shinde *et al.*, 1995).

Foliar Symptomology

Leaves of *C. dichotoma* and *M. alba* started developing symptoms of visible injury at 20% fly ash concentrations in soil which further increased with 25% and 30% fly ash levels. Similar reports of toxicity symptoms were reported on maize leaves when 35% and 50% fly ash treatments were given (Wong *et al.*, 1996). The injury due to fly ash was however similar to Boron toxicity symptoms. Similarly due to accumulation of B from fly ash amended soils, the leaves of *Agropyron elongatum* showed B toxicity symptoms in leaf tips at 35% and 50% ash amendment rates (Wong *et al.*, 1998). Fly ash addition in soil do not cause elemental toxicities to the plants except for B. Although visible B toxicity symptoms and elevated levels of plants were evident but there was no reduction in growth and biomass of *Hordeum vulgare* L. (Taylor and Schumann, 1988).

The reports for significant toxic levels of B in plant tissues leading to visible injury at early stage of development were also reported for *Bromus inermis*, *Medicago sativa* and *Hordeum vulgare* which became increasingly severe at later stages (Hammermeister *et al.*, 1998). Leaf margin curls and necrosis were markedly visible in Cucumber cv Poinsette plants grown on abandoned coal fly ash basin possibly due to toxic levels of B and Mn in leaf tissues (Dosskey and Adriano, 1993). Mo deficiency symptoms were also reported on fly ash enrichment in soils at very high concentrations (Barkowski and Beresniecwicz, 1989).

Fly ash through air also resulted in development of injury symptoms on leaves of *Azadirachta indica*, *Ficus glomerata* and *Nerium oleander* including interveinal and tip chlorosis and necrosis (Satyariaryana and Pushaplatha, 1991). However if exposure of fly ash and SO_2 was applied simultaneously, a specific mosaic chlorosis symptom developed on cucumber foliage (Tung *et al.*, 1995). The histopathology of this symptom revealed an unusual pattern of tissue abnormally initiated from the collapse of the anticlinal cell walls of the upper epidermal cells and progressed along cell walls into mesophyll tissues. Cellular injury at the lower spongy tissues was also observed. A significant elevation of sulfate - sulfur in the leaves treated with a combination of SO_2 and fly ash was demonstrated. It is suggested that fly ash deposit on the leaf surface play a catalytic role in oxidizing atmospheric SO_2 continuously to form sulfuric acid in situ. This acid could infiltrate into mesophyll cells thereby damaging cells and tissues enroute.

Wong *et al.* (1996) conducted a pot experiment to investigate the boron (B) bioavailability to maize seedlings in coal fly ash amended soil under greenhouse conditions. B content of plant tissue increased according to the rate of ash amendment and at high ash application rates (more than 70%) B was accumulated in leaf tissues. Boron toxicity symptoms in leaf margins were observed in 30% and. 50% ash amended soil.

Large addition of fly ash and compost high in B can also result in the B toxicity in crops (Gupta *et al.*, 1973; Purues and Mac Kenzie, 1973; Adriano *et al.*, 1978). The B toxicity symptoms are similar for most plants. Because of the slow transport of B in the plants, the symptoms generally appear on the older leaves and consist of marginal and tip chlorosis, which is quickly followed by necrosis (Shorrocks, 1974; Gupta, 1993; Gupta and Gupta, 1998).

Visible symptoms of Mn toxicity include stunted growth with general chlorosis, necrotic leaf spots, white flaking and leaf tip burn (Vlamis and Williams, 1964; Berg, 1978; Moraghan, 1979; Keisling *et al.*, 1984). Leaf necrosis may have relation with Vanadium in fly ash. The validate ion acts as inhibitor to the sodium pump and negatively affects many enzymes (Iqbal *et al.*, 2000).

These results and observations however contradict with the investigations of Pandey *et al.* (1994) for *Helianthus annuus* plant where they showed that the addition of fly ash to soil promoted plant growth with no visible symptoms of either nutrient deficiency or phytotoxicity. Similar results have been reported with corn (Moliner and Street, 1982), soyabean (Mishra and Shukla, 1986) and *Brassica chinensis* and *B. parachinensis* (Wong and Wong, 1990) Boron toxicity symptoms on leaves of barley were observed for various fly ash treatments (Sale *et al.*, 1996). Symptoms began as a white leaf tip bordered by a necrotic edge appeared on the leaf. The necrosis enlarged and advanced town, the leaf margins.

Random areas on the leaf veins became chlorotic then enlarged to form necrotic spots that overlapped into interveinal areas. The leaves with 6.25% treatment had slightly less affected leaf area and stayed green longer than leaves in other treatments. The soil in this experiment may have been deficient in Cu and or B for healthy plant growth, according to the criteria of Kabata Pendias and Pendias (1992). Adding 6.25% fly ash to the soil increased plants concentration of these elements to those considered normal for Cu but toxic for B. The young leaf tips of plants grown on all treatments turned white due to Cu deficiency symptoms. Concentrations of both B and Se increased steadily as the rate of fly ash added to the soil increased. Boron toxicity symptoms became evident with the addition of 6.25% fly ash, the lowest rate used. Reduced photosynthetic area caused by the necrotic and chlorotic lesions was hence attributed to B toxicity.

Foliar Morphology

In the present investigations all growth parameters viz. leaf area, leaf length and leaf width declined at high concentrations of fly ash (20%, 25% and 30%) but increased significantly with 5% fly ash for C. *dichotoma* while 5% and 10% fly ash for *M. alba.* The reduced leaf area in the polluted environment may indicate not only a retarded growth but sometimes also a morphological adaptation to the stressed environment (Nafees, 1995; Sharma and Sharma, 1996. Reduction in leaf area normally occurs with increasing concentrations of metals in the fly ash. Heavy metals in growth media can function as stressors causing physiological constraints that decrease plant vigor and inhibit plant growth (Saba *et al.,* 2000). The growth inhibition could also be the consequence of insufficient availability of photosynthetic assimilates which in turn could be ascribed to the reduction in the rate of photosynthesis as the heavy metals decrease photosynthetic rate (Setia *et al.,* 1993).

Fly ash increased leaf area of sunflower with the best results given by 0.5 kg fly ash, however as the dose of fly ash in soil increased the leaf area was lowered (Pandey *et al.,* 1994. The soil pH has been reported to increase with the increasing rate of fly ash. Similar reports of increase in leaf area and other growth parameters on fly ash amended soils have been reported for

Lactuca sativa L. (Srivastava *et al.*, 1995). Growth of lettuce seedlings was highest at 10% fly ash level as compared with the control however it was adversely affected by higher rates of fly ash (20% and 30%). Similarly in a pot experiment conducted to study the effect of fly ash on growth of wheat, data demonstrated that 50% fly ash applied to soil increased leaf number and leaf area (Tripathy and Sahu, 1997). Same results of boosting growth with various combinations of fly ash and soils have been reported (Mishra *et al.*, 1996; Singh *et al.*, 1996).

Increase in fly ash concentrations in the soil correspondingly increase the availability of carbonates, bicarbonates, sulfates, chlorides. B, P, K, Ca, Mg, Mn, Cu and Zn in the soil. The porosity, water holding capacity, pH, conductivity and cation exchange capacity also increased progressively in the fly ash amended soils (Khan *et al.*, 1997).

Increase in leaf length of *Brassica parachinensis*, and *B. chinensis* grown on fly ash amended soil was recorded with 3% and 6% treatments however high concentrations of 12%, 30% and 50'%) reduced the length of first leaves and all growth parameters (Wong and Wong, 1989). Fly ash samples contain a high content of various trace elements including Cd, Pb and other metals such as Co, Cu, Fe, Mn, Mo, Ni and Zn. Adding fly ash raises the electrical conductivity and pH of amended soils. The increase in electrical conductivity generally limits the availability of soil water because of the high osmotic pressure and the increased pH alters the availability of trace elements to plants. However the high contents of B, Mn and Zn in fly ash causes toxic levels resulting in growth suppression implying a limitation to successful vegetation establishment at very high concentrations of fly ash in soil though at low concentration the deficiency of these elements of soil are corrected (Dosskey and Adriano, 1993) thereby resulting in a better plant growth.

Relative growth rate is a measure of plant productivity and expresses the efficiency of the plant as a producer of new material (Hunt, 1992). Relative growth rate and specific leaf area increase at low concentrations of fly ash show that it enhances the growth at low concentrations. Many chemical constituents of fly ash might benefit plant growth and could improve agronomic properties of the soil (Chang *et al.*, 1977) but only at low rates of application due to better utilization of the nutrients derived from the fly ash by plants (Pandey *et al.*, 1994).

The better growth and yield of plants is apparently due to the presence of utilizable plant nutrients viz. sulphate, P, K, Ca, B, Mg, Mn etc. The nutrients from fly ash have been reported as being beneficial to plants through soil application or foliar dusting (Mishra and Shukla, 1986). The deficiencies of soil in plant nutrients specially P, B and K are corrected by the addition of fly ash to the soil resulting in better growth and yield (Khan and Khan, 1994). Availability of Boron present in fly ash amended soil and its relation to plant growth has been reported extensively (Mulford and Martens, 1970).

Porosity is the air space between soil particle which is usually occupied by water when available (Chopra and Kanwar, 1982). Therefore the increase in WHC is due to greater space between the soil particles. Electrical conductivity is positively correlated with pH and reflects the total concentration of soluble cations and anions (US Salinity Lab Staff, 1954; Elseewi *et al.*, 1978). Thus it is quantatively related to the concentration of salts. The higher values of EC have adverse effects on crop production. The increase in soil B content also cause decrease in plant growth and yield (Francois, 1984). However an increase in the availablity of P and other nutrients by fly ash provides a high degree of resistance in plants against toxicity of metals and their accumulation in foliage (Bardley *et al.*, 1981).

The gradual decline in the measured parameter at very high fly ash doses occurs probably due to salanity caused by the higher levels or sulphates, chlorides, carbonates and bicarbonates in the amended soils, It has been, demonstrated that foliar growth of H*ordeum vulgare* reduced under the synergistic effect of Cu and Zn (Brackett and Davis, 1978). Excessive uptake of these elements and their subsequent accumulation in the plants may account for the reduced growth and yield (Aitken and Bell, 1985). Some other toxic compounds viz dibenzofuran and dibenzo - p - dioxine mixtures (Helder *et al.*, 1982; Sawyer *et al.*, 1983) and metals viz. nickel, arsenic, cadmium, chromium, lead selenium, zinc, copper etc. (Mayer, 1981; Hale *et al.*, 1985; Wadge and Hutton. 1987;

Carlson and Adriano, 1993) are reported to occur in fly ash which might contribute towards the poor growth of plants at higher levels of fly ash.

As fly ash lacks nitrogen, its application especially at higher concentrations results in severe deficiency of nitrogen in soil as well as in the plant tissues which has been an important factor responsible for the suppressed growth and yield (Khan and Khan, 1996). The toxic effects of applying fly ash in soil also include a decrease in microbial activity. Application of ash at high rates inhibits the soil microbial respiration (Wong and Wong, 1986)

The application of alkaline fly ash to soil increases the soil pH markedly thereby reducing the extractability of trace elements that might result in reducing the availability of certain essential elements. This in turn can affect the plant growth through a deficiency toxicity complex (Hodgson and Holliday. 1966). On the other hand it has been well documented that low levels of trace elements in aqueous solution can cause inhibition and delay in seed germination (Adriano *et al.*, 1973) and trace elements in fly ash treated soil may potentially inhibit early plant growth (Vollmer e*t al.*, 1982).

However the increase in leaf area of fly ash dusted *Zea mays* L. and *Glycine max* L. plants showed an increase at low dusting rates while a significant reduction in leaf area and plant growth was recorded at the highest dusting rates (Mishra and Shukla, 1986). Fly ash dusting resulted in substantially elevated concentrations of Boron in plants that were earlier deficient in B (Chapman, 1965). Boron was absorbed through the leaf surface through intact leaf cuticles and stomata (Murray, 1984). Rohrman (1971) reported that fly ash consists of <10% water soluble components and there are reports of increased availability of B in fly ash amended soil (Martens, 1971). Reduction in plant growth at the high dusting rates may be attributed chiefly to the excessive uptake and accumulation of B and salinity caused by excessive soluble salts on the leaf surface. Mulchi and Armbruster (1981) have reported harmful effects of saline aerosol deposition on field grown maize and soyabean.

Photosynthetic Response

Photosynthesis is the ultimate physiological limitation to plant growth and development. The photosynthetic capacity of individual leaf is one factor that determines the growth and metabolic status of plants. The increased photosynthetic potential in leaves is also a possible approach in improving the plant yield (Bhatia *et al.*, 1996).

The photosynthetic driving force consists of ATP and NADPH but from these two components ATP is more dynamic upon the influence of external actors (Siedlecka *et al.*, 1997). It is known that photosynthetic efficiency depends on leaf area, chlorophyll contents and the stomatal response to the environment (Dwyer *et al.*, 1989; Ashraf *et al.*, 1995; Ayari *et al.*, 2000). The role of chlorophyll in photosynthesis is well established but the relationship between chlorophyll contents and the rate of photosynthesis is equivocal. The reduction in leaf area may be an additional factor responsible for decline in photosynthesis (Joshi *et al.*, 1993). This idea gets support from the work of Makino *et al.*, 1983 who reported that the loss of chlorophyll is not correlated to that of photosynthesis. The net rate of photosynthesis and stomatal conductance was reduced at high concentrations of fly ash. In the present study, it has been found to be increased significantly at low concentrations for both the species. Similar observations of increased photosynthetic rate has been reported for *Phaseolus vulgaris* grown on fly ash amended soils than control soil. Bazzag *et al.*, 1974 indicated that inhibition of whole plants, net photosynthesis was dependent on stomatal closure. Stomatal opening is related to CO_2 concentration which controls rate of photosynthesis. According to Heath and Packer (1968) closure of stomata results from an increase in CO_2 concentrations within the leaf. Internal CO_2 concentration in present investigations increased at very high fly ash doses in soil but at low doses of fly ash it also reduced. Nitrogen is the component of chlorophyll and photosynthetic enzymes and hence can influence photosynthesis greatly. The effects of heavy metals on photosynthesis have usually been studied using the isolated chloroplasts (Baszynski *et al.*, 1982). In general the heavy metals preferentially inhibit photosystem II (PS II) activity (Sandmann and Roger, 1980). The increase or decrease in photosynthetic rate with high doses of heavy metals appears to be associated with the change in total chlorophyll contents as influenced by presence of heavy metals. The extent of increment or decline in chlorophyll contents is dependent however on the concentrations of the heavy metals in soil (Setia *et al.*, 1993; Moya *et al.*, 1995).

Net assimilation rate (NAR) is a function not only of the photosynthetic rate but also of the rate of transpiration (Ledig 1974). The increase in NAR of fly ash treated plants than control plants may be due to increased leaf area of treated plants which helped in more carbon fixation, resulting in increased relative growth rate and photosynthetic rate. This shows better utilization of the nutrients derived from the fly ash by plants at the active phase of their growth (Pandey *et al.*, 1994).

The increase in soil pH after fly ash amendment results in improved plant growth attributed to the release of Ca, Na, Al and OH ions from fly ash (Hodgson *et al.*, 1982). It has been suggested that Ca forms $Ca(OH)_2$ in contact with water and being the major elements of fly ash it contributes maximum to the alkalinity of the amended soil. Later CO_2 is absorbed by $Ca(OH)_2$ leading to the precipitation of $CaCO_3$ (Wong and Wong, 1990). Over application of fly ash results in phytotoxicity due to very high concentrations of elements in fly ash however at low doses the metals remain well below the threshold limits resulting in improved plant growth and status (Singh and Yunus, 2000).

Green Pigments in Leaves

Chlorophyll is synthesized throughout a plant life and effects the overall status of growth and photosynthetic productivity of the plants (Lebedey *et al.*, 1992; Jain and Gadre, 1998). Chlorophyll loss is often assumed to be a symptom of stress injury (Hortensteiner, 1999; Gibon *et al.*, 2000). Since the active production of chlorophyll occurs during the growth of new leaves hence the effect of metals seems to be particularly pronounced during this development stage. Therefore the inhibition in synthesis of chlorophyll during early growth stage seems to represent the primary event due to heavy metal toxicity (Baszynski *et al.*, 1980). In the present study, chlorophyll and carotenoid contents showed a reduction at high doses of fly ash though a significant increase was reported with low fly ash treatments in soil for both the investigated species. The change in chlorophyll synthesis ultimately leads to changes in chloroplast development and photosynthetic processes (Stobart *et al.*, 1985).

The alteration in the ratio of chlorophyll *a* and chlorophyll *b* may be due to a partial conversion of the former into the latter as most evidences favour the view that chlorophyll *b* is synthesized from chlorophyll *a* (Keshan and Mukherji, 1992). The change in the levels of these pigments could be due to modification in their synthesis and/or its enhanced degradation leading to progressive change in its photosynthetic capacity (Mosguera and Guerrero, 1996).

Determination of chlorophyll content of plants is often acomplished to assess the impact of environmental stress as the pigment content is linked to the visual symptoms and photosynthetic plant productivity (Parekh *et al.*, 1990). At various fly ash rates the carotenoid content in leaves of *Phaseolus vulgaris* showed a considerable increase (Merakchiiska *et al.*, 1995). Similar enhancement in pigment contents of plants growing on low doses of fly ash in soil due to increased availability of heavy metals has been reported extensively, though at very high rates of fly ash there, had been adverse affects on leaf pigment contents (Srivastava *et al.*, 1995). This response is due to correction of B deficiency by fly ash however at high concentration of fly ash in soil there has been reduction in pigment content chiefly due to over accumulation of B (Mishra and Shukla, 1986).

The decline of chlorophyll content observed in various plant system in response to heavy metals is used as a marker to identify metal toxicity (Hampp *et al.*, 1976; Lamoreaux and Chaney, 1978). The decline in the level of pigments by the heavy metal suggests that these metals interfere with pigment metabolism (Prashad and Prashad, 1987; Bhattacharya and Choudhari, 1994). Imparing of chlorophyll synthesis by pressure of heavy metals may also be due to interference with the synthesis of proteins, the structural components of chloroplast (Keshan and Mukherji, 1992). The treatment with heavy metals presumably blocks the activity of the enzyme proteins responsible for chlorophyll biogenesis (Nag *et al.*, 1981). The reduction in photosynthetic area and inhibition of photosynthetic rate initiates the reduction in chlorophyll content (Setia *et al.*, 1993).

Application of unweathered fly ash on land could change the soil pH and subsequently induce the reduction of certain essential elements (Fe and Mn) available to plants (Wong and Wong, 1989). As a result a decrease in pigment at high doses of fly ash might have checked the uptake of essential micronutrients such as Fe and Mn required during the process of chlorophyll synthesis Reduction in total chlorophyll with high concentrations can also be attributed to the cumulative effect of highly soluble salt content and alkalinity which might have affected the specificity of enzymes and co - enzymes that play direct/indirect role in the formation and synthesis of pigments (Srivastava *et al* 1995). However the increase in photosynthetic pigments with 10% fly ash amended soil may be attributed to the availability of sonic useful elements/metals to the plants.

The addition of fly ash in soil increases the porosity, water holding capacity, pH, conductivity, cation exchange capacity, sulfate, carbonate, bicarbonate, chloride, P, K, Ca, Mg, Mn, Cu, Zn and B concentrations of the medium resulting in enhanced leaf carotenoid and chlorophyll concentrations at optimum dose of fly ash while an increase in dose retards the leaf pigment content (Khalil and Shah, 1979; Khan and Khan, 1996; Khan *et al.*, 1997).

At very high concentrations, apart from over accumulation of B there are evidences of presence of some other toxic compounds viz. dibenzofuran and dibenzo - p - dioxide mixtures (Helder *et al.*, 1982; Sawyer *et al.*, 1983) and metals viz. nickel, arsenic, chromium, lead, selenium, zinc, copper etc. (Wadge and Hutton, 1987) which might result in phytotoxicity at high concentrations of fly ash in soil (Deckock, 1956; Poison and Adams, 1970; Berg 1978; Shukla and Mishra, 1985).

The chlorophyll content of the fly ash dusted leaves increased significantly (Dubey *et al.*, 1982; Mishra and Shukla, 1985). This could be due to shading effect caused by fly ash deposition (Oblisami *et al.*, 1978). However this might not be the only factor because fly ash contains water - soluble salts (Rohrman, 1971) and leaf surfaces are known to absorb nutrients. This may thus be a cumulative effect. Reduction in chlorophyll content at high dusting rate is attributed to the alkalinity caused by excessive soluble salts on the leaf surface (Elseewi *et al.*, 1980) and also to increased foliar temperature which retards chlorophyll synthesis (Mark, 1963). Chlorophyll may be destroyed in the cells under a thick fly ash cover, as has been reported with reference to cement dust (Borka, 1980). It is also suggested that like ozone (Fletcher *et al.*, 1972; Adedipe *et al.*, 1973) or sulfur dioxide (Beckerson and Hofstra, 1979), coal particle pollution also causes induction of chlorophyllase or breakdown of chlorophyll to form phaeophytin (Mudd, 1975). Similarly photosynthesis was faster at low to moderate fly ash application but declined at increased dusting rate in the maize and soyabean crops (Mishra and Shukla, 1986), possibly because increased foliar temperature accelerates photosynthetic activity at low rates of fly ash application, where a thick fly ash layer interferes with light. In *Hibiscus abelmoschus*, cement dust was more damaging to chlorophyll than the coal dust, while fly ash slightly increased the chlorophyll content (Pawar *et al.*, 1982).

Total Soluble Protein

High protein accumulation in wheat leaves enhanced the protein content of grain (Maldonado *et al.*, 1996). Storage proteins are tissue specific. Change in the pattern and type of protein synthesized during the development and maturation of seeds may provide convenient points for manipulation of genes responsible for protein quantity and quality (Dhillon and Nainawatee, 1990). Changes in soluble protein contents may be associated to changes in photosynthetic rate. During leaf senescence, total soluble protein content charged dramatically which may be attributed to dissolution of insoluble proteins into soluble proteins (Fu *et al.*, 2000).

Plants synthesize metal binding peptides when exposed to metal stress (Steffens, 1990). The total soluble protein content in present study declined with increase in fly ash content of soil. Nevertheless it increased in fly ash treated plants at low doses at all stages of growth in both the species. The higher protein content in stressed saplings may be due to the synthesis of additional stress shock protein to alleviate stress (Vierling, 1991; Prasad and Prasad,1987). The increased protein content was also thought to be the consequence of (a) stimulated gene expression and de novo synthesis (b) a increased rate of degradation of proteins in the apoplast (c) altered cell wall structure facilitating extraction (d) decreased insolubilization of apoplastic protein in the cell wall fabric. Evidence is present that enhanced protein synthesis is of particular importance for the phenomenon of the heavy metal induced increase in apoplastic proteins (Blinda *et al.*, 1997).

Fly ash, an amorphous ferro alumino silicate mineral with major matrix elements being Si, B, Zn, Cd, Pb, Al, Cr etc. (Martens *et al.*, 1970), is known for its utility for the improvement of yield and quality of various crops. The protein content of green gram increased with increasing doses of fly ash due to increased nutrient uptake from fly ash amended soil. The increase in protein content was attributed to higher test weights due to better development of seeds as a result cf supply of essential nutrients from fly ash amended soil (Warambhe *et al.*, 1992; Bhaisare *et al.*, 2000).

Activity of Nitrate Reductase and Nitrate Content

Nitrate reductase is the enzyme that regulates the first step of nitrogen assimilation (Abrol *et al.*, 1976). Its activity is directly correlated to nitrate concentrations in leaves. When NR activity is high it will reduce a great amount of nitrate that will be utilized further in protein synthesis. Thus nitrate concentration in these leaves will be low than in those plants with low NR activity because the latter can reduce less amount of nitrate.

In the present investigation, NR activity increase at low fly ash doses in soil but got reduced as the concentration of fly ash increased in both the investigated species. The leaf nitrate content however decreased in both the species with increasing fly ash rates in soil. Similar reports of increase and decrease in leaf NR activity at low and high fly ash doses respectively were reported in a greenhouse trial with *Phaseolus vulgaris* (Kamenova *et al.*, 1995). But increased NO_3 concentration in plant tissues with various fly ash amendments due to elevated levels of NO_3 in soil have been reported in sorgrass (Sajwan *et al.*, 1996). The concentration of NO_3 and various other heavy metals from fly ash increase in soil as compared to control resulting in the release of nutrients including NO_3 into the system. Plants absorbed larger amounts of water soluble nutrients when available in the amended soils. This may partly explain the high levels of certain elements in the plants grown in soil amended with fly ash. Although heavy metals such as Pb, Cd, Ni and Cr were relatively high in the plants grown in soils amended with fly ash but they remained well below the threshold levels.

Nitrogen being a component of plants promotes growth and development of increasing nitrogenous metabolites (Mishra *et al.*, 1994) and acts as a regulator of hormonal level as well (Singh *et al.*, 2000). Nitrogen metabolism is an important metabolic process that has a special relevance with plant growth. Nitrate reductase is usually considered to perform the rate limiting step of the nitrate assimilation pathway owing to the low turnover of NR compared with the other enzymes in the pathway and to play a central role in regulation (Marwaha, 1998; Kaiser *et al.*, 1999; Tischner, 2000). The activity of NR has often been correlated to the nitrogen status of the plants (Mishra and Srivastava 1983; Dabas and Singh, 1995). In addition NR is very convenient genetic marker that has the potential to be selected or counter selected in many organisms (Rouze and Caboche, 1992). The leaf nitrate content is probably the most decisive factor controlling NR expression and activity in shoots and roots (Abdel Bak *et al.*, 2000).

In many plant species, the leaves are primary sites of nitrate reduction. Leaf vacoules have been reported as the major storage pool for nitrate. The exposure of plants to high fly ash doses may lead to vacuolar accumulation of heavy metals (Rauser and Ackerly, 1987). Thus it seems that the metal toxicity involves reduction in metabolization of the endogenous nitrate (Vyas and Puranik, 1993). A decrease in leaf NR activity may be attributed to a restricted translocation of the metal to the site of enzyme action in the leaves (Bharti and Singh, 1993).

The strong affinity of various heavy metals to side chain ligands of proteins strongly suggest that enzyme activities are one of the first metabolic processes to be interfered by heavy metals (Vallace and Ulmer, 1972; Hampp *et al.*, 1976). The inhibition rates of plant enzymes by heavy metals are dependent on the concentration of metals (Weigel *et al.*, 1990; Hemandez, *et al.*,

1997; Gouia *et al.*, 2000). Several authors have also found positive correlation between leaf NR activity and leaf protein accumulation (Maldonado *et al.*, 1996).

Content of Soluble Sugar

Carbohydrates, the direct products of photosynthesis are the basic energy storing compounds that accumulate as reserve food (Kramer, 1969). The state of carbohydrate could be a good pointer to the physiological process in plants. *Beta vulgaris* grown in fly ash amended soil revealed the application of low amount of fly ash favored plant growth and improved yield (Singh *et al.*, 1994). The increase in sugar content at low doses of fly ash amendment enhances the possible use of fly ash in crop yield improvement however higher concentrations of fly ash showed adverse effect. The increase in Cd, Cu, Fe, Mn, Ni and Pb content in plant tissues was also reported though their levels remained well below the threshold limits and thus suitable for human consumption. The soil pH and electrical conductivity was also improved for increased plant, growth and yield. Similarly the leaves of *Datura innoxia* when dusted with fly ash led to an increase in levels of sugars and high activity of oxidative enzymes (Satyanaryana *et al.*, 1988).

The above reports are almost similar to the results of present investigation where at low doses of fly ash the leaf sugar content increased though at high concentrations of fly ash it decreased significantly at all stages of growth in both the investigated plant species. The plant uptake of K, Mn and Cu increases with increasing concentrations of fly ash resulting generally in an increased plant growth and yield (Ghuman *et al.*, 1994) however, over application of fly ash results in phytotoxicity, B being the most problematic [Elseewi *et al.*, 1981; Zwick *et al.*, 1984).

Hydrogen Ion Concentration of Leaves and Soil

Environmental pH is one of the most significant factors affecting the eco-physiological processes. The solubility of metals in soil solution is pH dependent. It increases with the increase in pH, hence the toxicity of metals may be pH regulated (Youssef and Chino, 1991; Pawlik *et al.*, 1993; Nwosu *et al.*, 1995). The physiology of uptake and translocation of the metals in the plants is the function of the variation in soil pH (Lagerweff, 1971; Wong and Wong, 1989). In the present study, the soil pH increased with fly ash enrichment of soil than the control soil. Similarly the pH of leaves increased as doses of fly ash in soil increased for both the studied species. The pH in soil and leaves exhibited a positive correlation.

Various observations of increase in soil pH due to fly ash addition have been reported by various investigators (Adriano *et al.* 1982; Korcak, 1984; Cervelli *et al.*, 1986; Taylor and Schumann, 1988; Borkowski and Beresniewicz, 1989; Kukier *et al.*, 1994; Pandey *et al.*, 1994; Shinde *et al.*, 1995; Menzies and Aitken, 1996; Bharti *et al.*, 1999). Fly ash ranges from extremely basic to acidic usually in the range of 6.5 - 11 pH and thus varies greatly in its neutralizing capacity depending on the source of coal (Babcock, 1972: Fail and Wochok, 1977). The physical, chemical and mineralogical characteristics of fly ash depend on a variety of factors including the composition of the parent coal, combustion conditions, the efficiency and type of emission control devices and the disposal methods used (Van Hook, 1979; Adriano *et al.*, 1980). Consequently it is difficult to generalize the composition of ashes or their behaviour in the environment. The pH of some alkaline ashes can exceed 12 and can cause deficiencies of essential elements such as Cu, Fe, Mn and Zn. Alkaline ash can cause increased accumulation of some non-essential trace elements such as As, *Se* and V whose solubilites are pH dependent (Page *et al.*, 1978). Fly ash is high in oxides of Ca and Mg that contribute partly to increase in soil pH (Singh *et al.*, 1994; Khan *et al.*, 1996).

Foliar and Soil Nitrogen Content

Fly ash addition to soil in the present study caused no significant change in soil nitrogen content. While Khan and Khan, 1996 reported that fly ash addition to soil caused significant reduction in nitrogen concentration. From chemical viewpoint all naturally existing trace elements enrich fly ash (Kaakien *et al.*, 1975; Klein *et al.*, 1975) however ashes are typically low in N due to volatilization during combustion (Fail and Wochok, 1977; Adriano *et al.*, 1980; Bradshaw and Chadwick, 1980; Nass *et al.*, 1993). Lack of N in fly ash limits partially the colonization of ash deposits by microrganisms (Cope, 1962; Rippon and Wood, 1975).

However application of nitrogen fertilizers to fly ash significantly increased the contents of N in soil as well as plant tissues grown on fly ash amended soils at all growth stages though over application of fly ash decreased the N content and its uptake (Singh *et al.*, 1992; Malewar, 1999). There have been reports where leaf N content was not significantly affected by fly ash treatment. (Korcak, 1985). Fly ash amendments attributed deficiency of nitrogen in plant tissues though there has been a significant enrichment of all other elements. Decrease in contents of available N might be due to lack of this nutrient in available form present in fly ash (Menon *et al.*, 1993). The result however in contrast with the result of Kene *et al.*, 1991 where the increase in N content and uptake was recorded at low concentrations of fly ash application though at higher doses of fly ash addition, there was interference in N nutrition of crops (Singh and Singh, 1986). Similarly in the present investigations the leaf N content increased at low percentage of fly ash in soil though declined significantly at higher doses of fly ash in both the species. Nitrogen is associated with the organic component of the coal and is expected to be lost during the burning of coal in the furnaces thereby leading to N deficiency in plants grown on fly ash (Lepp *et al.*, 1977). As fly ash lacks nitrogen its application especially at higher concentrations resulted in severe deficiency of N in soil as well as in plant tissues (Carlson and Adriano, 1993; Khan *et al.*, 1997.

Because fly ash contains appreciable quantities of salts there might be an increase in the osmotic potential ultimately restricting the uptake of major nutrients (Singh and Singh, 1986; Shinde *et al.*, 1995). At low levels of fly ash amendment in soil the N availability increases resulting in high N uptake by plants though at higher application rates the decrease in nitrogen content of plant tissues is related to the presence of salts in fly ash. Since the fly ash contain salts which might increase the osmotic potential and finally inhibit, the absorbtion of nutrients resulting in poor uptake of nutrients (Cherenkova, 1977). The deficiency of N can easily be alleviated by addition of N fertilizers even at very low rates (Hodgson and Townsend, 1973; Townsend and Gillham, 1975; Bradshaw and Chadwick, 1980).

Sulphur Content in Soil and Leaf Tissues

Sulphur content in soil and leaves exhibited a non - significant decreasing trend with increasing treatments of fly ash in soil in the present investigation. Fly ash, comprising of much minute glass like particles, is rich in a variety of elements such as As, B, Ca, Mo, S, Se and Sr (Page *et al.*, 1978; USEPA, 1988; Singh and Yunus, 2000). The S content in fly ash varies from low to very high depending on the S content of parent coal (Swanson, 1972; Block and Dams, 1975).

The present findings contradicts with the earlier investigations of Eaton (1965) where the total S content of plants not receiving fly ash ranged from 0.057 to 0.066% and falls within a deficiency range. Addition of upto 8% fly ash to the soil increased S concentrations in soil and plant tissues as well Elseewi *et al.*, 1978). Plants grown on sulfur deficient soils respond favourably to incorporation of fly ash to the soil in response to increased concentration of available S thereby leading to increased S content in plant tissues. Fly ash is thus a potential inexpensive source of S particularly in acid soils where deficiencies of S are common.

Foliar Anatomy in Relation to Epidermal Responses

The adaptive responses of stomata have long been of interest to physiologists (Croxdale, 2000). With decreased stomatal size, the stomatal conductance is likely to decline. Kastori *et al.* (1992) while working on water relations in *Helianthus annuus* has observed an increase in number of stomata per leaf area and a decrease in size of the stomata under excess of Pb, Cd, Cu and Zn (Lindon and Henriques, 1998). The excess concentrations of these elements significantly affect plant water status thus causing a water deficit and the subsequent metabolic changes in the plants (Tomar *et al.*, 2000) In the present study also stomatal density decreased with increasing fly ash dose on both adaxial and abaxial surfaces of leaves in both the species while at low concentrations it increased significantly. However the stomatal size and the stomatal pore size experienced an irregular trend of increase and decrease with increasing fly ash doses.

According to Wright and Woolhouse (1977) the apparent increase in number of stomata is not due to the increased stomatal formation but rather due to reduction in size of neighbouring cells brought about by water deficit caused by heavy metals. In *Phaseolus vulgaris* the direct effect of heavy metals on stomatal opening may be caused by the alteration of K fluxes through membranes (Rauser and Dumbroff, 1981). The unusual thickening of the wall of guard cells in treated plants of *Vigna radiata* seems to keep the stomata partially closed (Klein *et al.*, 1996). Therefore the closure of stomata has been a result of excessive metal concentrations of fly ash amended soils.

With reference to fly ash through air pollution there have been extensive reports on decreased stomatal frequency and size (Tietzeova and Makusova, 1981; Mishra, 1982). Rates of gaseous exchange were higher in the fly ash dusted leaves and was attributed to particles on the leaf surface interfering with the opening and closing mechanism of the stomata and thus blocked the closing of the stomata (Krajickova and Mejstrick, 1984). Scanning electron microscopy and X - ray analysis confirmed the presence of elements from fly ash settled on the stomata of foliage. Particles of fly ash viz. Si, Al, Mn and Fe were present in 98% of foliage stomata of *Picea abies* and *Fagus sylvatica* (Mankovska, 1997).

However inconsistent effects of fly ash on plant growth have been reported from different regions of the world (Adriano *et al.*, 1973; Vollmer *et al.*, 1982). The discrepancies are possibly due to the differences in the heterogeneity of the physical, mineralogical and chemical properties of fly ash which depend on the composition of the parent coal, combustion conditions, efficiency of the emission control devices, storage and handling of fly ash and differences in climatic conditions (Adriano *et al.*, 1980; Ramsden and Shibaoka, 1982). The physical and chemical properties of the recipient soils and the type of plant species to be grown further complicate the problem. Therefore favorable plants response will be dependent on a proper combination of soil types, plant species and regional practices (Wong and Wong, 1987).

6

Concluding the Results

The present investigations undertaken to assess the impact of fly ash amended soil on foliar features of *Cordia dichotoma* Forst f. and *Morus alba* Linn. established that fly ash exerts a profound effect on both the species. The tree saplings subjected to different fly ash concentrations in soil were compared with themselves and against control unamended soil. Following are major conclusions derived from the present investigations.

- Foliar injury symptoms appeared on both the tree species at higher concentrations (20%, 25% and 30%) of fly ash in soil while at low concentration (10%) there were no visible injury symptoms.

- The Leaf area, leaf length and leaf width declined significantly at higher concentrations (20%, 25% and 30%) of fly ash in soil while at 5% and 10% they showed a significant increase in *M. alba* while in *C. dichotoma* there was a significant decrease at all concentrations of fly ash except at 5% at all stages of plant growth.

- Photosynthetic pigments viz. total chlorophyll, chlorophyll *a*, chlorophyll *b* and carotenoid contents showed an increase at 5% in *C. dichotoma* and 5% and 10% in *M. alba* and thereby exhibited a linear dose dependent reduction with maximum reduction at 30% of fly ash in soil at all the developmental stages of are species.

- Total soluble protein in *C. dichotoma* leaves exhibited an increase at 5% of application while in *M. alba* leaves an increase was recorded at 5% and 10% thereafter a reduction at high rates of fly ash was recorded as against control plants.

• Foliar nitrate content increased and nitrate reductase activity decreased with the influence of 5% fly ash application in *C. dichotoma* and at 5% and *10%* in *M. alba*.

• Soluble sugar content of leaves increased at each developmental stages of both the species at 5% in *C. dichotoma* and up to 10% in *M. alba* thereby showing a significant dose dependent decrease.

• The leaf pH showed a direct positive correlation with plant age and dose of fly ash in soil in both the tree species as compared to control unamended soil.

• The nitrogen content of the leaves of *C. dichotoma* exhibited an increase at 5% of fly ash application while in *M. alba* leaves the increase was recorded up to 10%. Thereafter at higher doses of fly ash, there was an insignificant reduction in nitrogen content in both the species.

• The sulphur content of leaves was least affected by fly ash at any concentration in both the species.

• The rate of photosynthesis and stomatal conductance were appreciably greater at lower concentrations of fly ash particularly at 5% in *C. dichotoma* and up to 10% in *M. alba*. However, the high concentrations up to 30% of fly ash resulted in a marked decline in both the species at each stage of plant development

• The intercellular CO_2 concentration was always high at higher doses of fly ash in soil in both the species, but it declined at 5% and up to 10% in *C. dichotoma* and *M. alba* respectively.

• Stomatal index of abaxial surface reduced maximally in *C. dichotoma* and *M. alba* at 30% concentration of fly ash, however maximum increase for *C. dichotoma* and *M. alba* was recorded at 5% and 10% fly ash levels respectively.

• Stomatal length and length of pore aperture on the abaxial surface of *C. dichotoma* and *M. alba* declined most significantly with 30% fly ash treatment. The maximum increase in size of stomata and stoma was recorded at 5% and 10% doses of fly ash in *C. dichotoma* and *M. alba* respectively.

• On the abaxial leaf surface, the stomatal width of *C. dichotoma* was maximum at 10% fly ash and minimum at 25% fly ash but in

M. alba, it increased with increasing fly ash concentrations as against control plant.

• The width of stomatal aperture on the abaxial epidermis was maximum in size at 5% treatment of fly ash for *C. dichotoma* and at 10% level of fly ash for *M. alba.* The minimum size was recorded at 30% for both the species.

• Stomatal index of adaxial epidermis of *C. dichotoma* reduced at all fly ash treatments except at 5% dose, where stomatal index increased appreciably.

• On the adaxial surface of *C. dichotoma,* the maximum increase and the maximum decrease in stomatal length was recorded with 5% and 30% fly ash treatments respectively.

• The stomatal pore length on the adaxial epidermis of *C. dichotoma* declined linearly with increasing concentration of fly ash in the soil, the maximum decline being at 30%.

• Stomatal width increased with the increasing fly ash concentrations on adaxial epidermis of *C. dichotoma,* however the stomatal pore width declined with the increase in fly ash level of soil in comparison to control test plants.

• The pH of soils treated with fly ash increased with the increasing fly ash dose.

• The S content of soil did not exhibit any marked significant change with the increasing fly ash treatment of soil.

• Soil Nitrogen increased as the concentrations of fly ash in soil increased, the maximum being at 30% and the lowest at 5% fly ash dose.

7

Summary: Utilization of Fly Ash

Environmental contamination of air, water, soil and food has become a threat to the existence of plants and animals. The pollution has especially affected the global agriculture due to severe impact on crop plants.

Human population growth has resulted in quite massive impacts on natural environment, particularly in the regions of industrial and urban developments in the western world. The massive urbanization and industrialization have resulted in greater fossil fuel combustion due to increase in energy demands affecting the long term availability of these non - renewable resources and causing major contamination of the earth's environment and changes in global climate. In addition the release of organic and inorganic pollutants resulting from a whole range of human activities threatens the future well being of terrestrial, fresh water and marine environments.

The thermal power plants cause a great deal of environmental contamination of air, water and soil; thereby causing a serious threat to the existence of diverse flora and fauna. The coal is the chief fuel driving the function of power plants. Increased use of coal for electric power generation results in large quantities of fly ash. Coal ash is basically an amorphous Ferro alumino silicate, which is high in elemental content. Basically all natural elements enrich fly ash as compared with the parent coal.

The fly ash thus produced has tremendous potential to be utilized for different applications. However, only about 10% of fly ash is being utilized and the rest is deposited in landfills or settling ponds. It is desirable to establish vegetation on the landfills for stabilization against wind and water erosion. Revegetating these areas is however difficult due to the presence of physical and chemical conditions in the ash that are deleterious to plant survival and growth. The most significant problem is that there may be considerable variation in the composition of coal ashes from different power plants, and even in separate samples of ash from the same generating plant. Therefore situations involving different ashes are sometimes difficult to compare. A second problem is that there is insufficient literature or the growth of plants on fly ash amended soils. Also soil properties vary considerably from one soil to the next, so the situations involving different soils are sometimes difficult to compare. Attempts have been made to use manure, hay, woodchips, sewage sludge etc to amend fly ash for successful vegetation establishment, although no perfect protocol could be formulated as yet because of differences in ash composition and the species planted.

Fly ash disposal and utilization is an important area of concern. The scenario with respect to fly ash management has undergone considerable improvement over the past few years. A beginning has been made towards enhanced facilitation for implementation of better management practices and appropriate technologies towards safe disposal and utilization of fly ash. Several areas of fly ash utilization include mine filling, construction of roads, embankments, raising of dykes, manufacture of general building components like bricks, blocks, tiles and its use in agriculture. Identification of crops and their cultivars capable of doing well in presence of fly ash particulates has become necessary for getting sustained yields under the changed scenario.

The present venture estimated the feasibility and ecological impact of growing trees on fly ash amended soil by finding out the rate of application of fly ash in soil at which both the species responded well with flourished and luxuriant growth. The leaf represents the status of plant growth and metabolic activities. Leaves are the primary sites of photosynthesis and the products of their metabolism fuel the growth of all plant tissues. Leaves are the most exposed organ of plants and affected most by the environmental stresses. The rate of photosynthesis is used as an early detector of the potential stress injury to plants. In the present research it was concluded that 5% and 10% of fly ash treatments to the soil were best - suited growth media for *Cordia dichotoma* Forst f. and *Morus alba* Linn. saplings respectively. But at a relatively higher dose of fly ash, undesirable phytotoxic effects were exhibited on both the investigated species, the maximum toxicity level being at 30%.

There have been several reports that fly ash produce undesirable effects on plants. The most frequently cited cause of these effects is B toxicity. Induced P deficiency, salt injury, poozolanic effects and heavy metals toxicity is also mentioned as possible mechanisms by which fly ash interrupts in plant growth. However, fly ash also has beneficial effects on plant development, which may be attributed to fly ash corrected nutrient deficiency in the soil. *Fly* ash also buffers the soil pH and supplements nutrient levels of soil usually at low doses.

Changes in soil characteristics and the cropping system in response to addition of fly ash particulate needs to be studied. The major factors limiting vegetation establishment on fly ash amended soils are: (1) lack of essential nutrients usually N and P (2) toxicity caused by high pH and high soluble salt concentrations (3) high concentration of Boron (B) and other potentially toxic elements (4) presence of compact and cemented layer in the ash, and (5) the sterile nature of fly ash.

The large amount of coal ash is generated each year and a great deal of research has to be conducted to identify and determine the feasibility of utilizing these wastes in agriculture and industry. Amendment of agriculture soil with fly ash can affect the soil and the chemical composition of plant, improve soil texture and increase water - holding capacity of soil. However, care is needed while using fly ash as a source of nutrients, since over application could result in phytotoxic level of B and sufficiently elevated levels of other elements including As, Mo and Se, thus posing a threat to animals consuming these tissues thereby leading to food chain risks. The future poses challenge to the scientists, technologists and engineers towards sound management of fly ash disposal and deposition technologies. The problem is however not due to lack of technical competence but due to lack of adoption and implementation of appropriate techniques of effective utilization of fly ash.

Bibliography

Abd-El Baki, G.K., Siefritz, F., Man, H.M., Weiner, H., Kaldenhoff, R. and Kaiser, W.M. (2000). Nitrate reductase in *Zea mays* L. Under salinity. Plant cell Environ., 23, 515-521.

Abrol, Y.P., Kaim, M.S. and Nair, T.V.R. (1976). Nitrogen assimilation, its mobilization and accumulation in wheat *(Triticum aestivum* L.) grains. Cereal Res. Comm., 4, 431-440.

Adedipe, N.O., Fletcher, R.A. and Ormrod, D.P. (1973). Ozone lesions in relation to senescence of attached and detached leaves of tobacco. Atoms. Environ., 7, 357-361.

Adriano, D.C. Chang, A.C. Pratt, P.E. and Sharpless, R. (1973). Effects of soil application of dairy manure on germination and emergence of some selected crops. J. Environ. Qual., 2, 396-399.

Adriano, D.C., Page, A.L., Elseewi, A.A. and Chang, A.C. (1982). Cadmium availability of Sudan grass grown on soils amended with sewage sludge and fly ash. J. Environ. Qual., 11, 197-203.

Adriano, D.C., Page, A.L., Elseewi, A.A., Chang, A.C. and Straughan, 1. (1980). Utilization and disposal of fly ash and other coal residues in terrestrial ecosystem: A review. J. Environ. Qual., 9, 333-344.

Adriano, D.C., Woodford, T.A. and Giravolo, T.G. (1978). Growth and elemental composition of corn and bean Seedlings as influenced by soil application of coal ash. J. Environ. Qual., 7, 416-421.

Ahmed A. and Alam, M.M. (1997). Utilization of fly ash and *Paecilomyees lilacinus* for the control of *Meloidogyne incognita.* Inter. J. Nemat., 7:2, 162-164.

Aitken K.L. and Bell L.C., (1985). Plant uptake and Phytotoxicity of boron in Australian fly ashes. Plant and Soil., 84, 245-57.

Anonymous (1962). The Wealth of India, Raw materials, Vol. VI, Publication and Information Directorate, CSIR, New Delhi.

Anonymous (1992). The Wealth of India, Raw materials, Vol.4 (Revised), Publication and Information Directorate, CSIR, New Delhi.

Anonymous (1997). Air Pollution on the rise despite checks. The Hindustan Times, New Delhi, March 15.

Anonymous (1997). Air Pollution on the rise despite checks. The Hindustan Times, New Delhi, November 2.

Anonymous (1998). The Observer of Business and Politics, 13th October. Anonymous (1999). The Business Line, December 22.

Arnon, D.I. (1949). Copper enzymes in isolated chloroplasts polyphenol oxidase in *Beta vulgaris.* Plant Physiol., 24, 1-15.

Arthur, M.A., Rubin, G., Schneider, R.E. and Weinstein, L.H. (1992). Uptake and accumulation of selenium by terrestrial plants growing on a coal fly ash landfill. Environ. Toxicol. Chem., 11:9, 1301-1306.

Arthur, M.A., Rubin, G., Woodbury, P.B., Schneider, R.E. and Weinstein, L.H. (1992). Uptake and accumulation of selenium by terrestrial plants growing on a coal fly ash landfill. Part 2, Forage and Root Crops. Environ. Toxicol. Chem., 11:9, 1289-1299.

Ashraf, M.Y., Khan, A.H. and Naqvi, S.S.M. (1995). Relationship of chlorophyll content and leaf area with grain yield in wheat genotypes. Indian J. Plant Physiol., 38, 162-613.

Ayari, O., Samson, G., Dorais, M., Boulanger, R. and Gosselin, A. (2000). Stomatal limitation of photosynthesis in winter production of green house tomato plants. Physiol. Plant, 110, 558-564.

Badhwar, (1944). In: The Wealth of India, Raw materials (1992), Vol. 4 (Revised), Publication and Information Directorate, CSIR, New Delhi.

Banerjee, M. and Deb, M. (1996). Potential of fly ash and Spirulina combination as a slow release fertilizer for rice field. *Cinetifica Jaloticabal,* 24:1, 55-62.

Banerjee, S.K. and Kashyap, M.K. (1998). Environmental impact of fly ash and its management. Adv. in Forest. Res. in India, 18, 31-53.

Bardley, R. Burt. A.J. and Read, D.J. (1981). Mycorrhizal infection and resistance to heavy metal toxicity in *Calluna vulgaris.* Nature, 292, 335-7.

Barkowski, J. and Beresinewicz, A. (1989). Lignite ash as a multi-component fertilizer for tomatoes in protected cultivation. Biuletyn *Warzywniczy,* Supl. II, 107-111.

Barnes, J.D., Balaguer, L., Manrique, E.. Elvira, S. and Davison, A.W. (1992). A reappraisal of the use of DMSO for the extraction and determination of chlorophylls a and b in lichens and higher plants. Environ. Exp. Bot., 32, 85-100.

Baszynski, T. Wajda, L., Krol, M., Wolinska, D., Krupa.Z. and Tukendorf, A. (1980). Photosynthetic activities of Cadmium treated tomato plats. Physiol. Plant, 48, 365-370.

Baszynski, T., Krol, M., Krupa, Z., Ruszkowska, M., Wojcieska, U. and Wolinska, D. (1982). Photosynthetic apparatus of Spinach exposed to excess copper. *Z. Pflanzenphysiol.* , 108, 385-395.

Bazzaz, F.A., Rolfe, G.L. and Carlson. R.W. (1974). Effect of Cadmium on photosynthesis and transpiration of excised leaves of Corn and Sunflower. Physiol. Plant, 32, 373-376

Beckerson, D.W. and Hofstra, G. (1979). Stomatal responses of white bean to ozone and sulfur dioxide singly or in combination. Atoms., Environ., 13, 533-535.

Belavady and Balasubramanium (1959). In: The Wealth of India, Raw materials (1962), Vol. VI, Publication and Information Directorate, CSIR, New Delhi.

Berg, W.A. (1978). Aluminium and Manganese toxicities in Acid Coal mines Wastes. In: G.T. Goodman and M.J. Chadwick (eds.). Environmental Management of Mine Wastes 141-150. The Netherlands. Barley, New Phytol., 81, 155-173.

Bernasconi, M.G., Calandra, R., Ciappelloni R. and Angelucci, M. (1999) Preliminary essay on fly ash and bottom ashes from power plants in Umbria. *Annali Della Facolta di Angraria, Unicersita Degli Studi di Perugia* 1996, Pub. 1999, 50: 47 - 54.

Bhaisare, B., Matte, D.B., Badole, W.P. and Deshmukh, A. (2000). Effect of fly ash on yield, uptake of nutrients and quality of green gram grown on vertisol. J. Soils and Crops, 10: 1, 122-124.

Bharti, B., Matte, D.B., Badole, W.P., Deshmukh, A. and Pillewan, S.

(1999) Effect if fly ash on physico-chemical properties of vertisol and yield of green gram. ,J. Soils and Crops, 9:2, 255-257.

Bhatia, V.S., Tiwari, S.P. and Joshi, O.P. (1996). Inter relationship of leaf Photosynthesis, specific leaf weight and leaf anatomical characters in Soyabean. Indian J. Plant Physiol., 1, 6-9.

Bhatt, D.K., Sanyal, P., Kumar, V., Varma, D.N. and Kumar, A. (1994). Development of fly ash brick plant : An R and D effort. Fert. News, 39:12, 51-57.

Bhattacharyya, M. and Choudhuri, M.A. (1994). Effect of lead and Cadmium on the biochemical changes in the leaves of terrestrial (Vigna) and aquatic (Hydrilla) plants under solution culture. Indian J. plant Physiol., 37, 99-103.

Bhumbla, D.K. (1992). Ameliorative effect of fly ashes. In : Dissertation-Abstracts International-B, Sciences and Engineering, 52:8, 3952B, 312 pp. Abstract of thesis, West Virginia University, U.S.A.

Blinda, A., Koch, B., Ramanjulu, S. and Dietz. K.J. (1997). De novo synthesis and accumulation of apoplastic proteins in leaves of heavy metal exposed barley seeplings. Plant cell Environ., 20, 969-981.

Block, E., and R. Dams. (1975). Inorganic Composition of Belgian Coals, and coal ashes. Environ. Sci. Technol.. 91, 146-150.

Borka, G. (1980). The effect of cement dust pollution on growth and metabolism of *Helianthus annus.* Environ. Pollut., 22, 75-79.

Borkowski, J. and Berensniewicz, A. (1989). Lignite ash as a multi component fertilizer for tomatoes in protected cultivation. *Biuletyn Warzywnicz,* II, 107-111.

Brackett, P.H.T. and Davis, P.D. (1978). New Phytol., 81, 155.

Bradford, M.M. (1976). A rapid and sensitive method for the quantification of microgram quantities of protein using principle of protein dye binding. Analyt. Biochem., 72, 248-259.

Bradshaw, A.D. and M.J. Chadwick, (1980). The restoration of land. Univ. of California Press. Berkeley, CA.

Brahmachari, K. Debnath, A. and Mondal, S.S. (1994). Effects of fly ash on yield and uptake in rice and its impact on soil properties. J. Interacademicia, 3:3-4,293-300.

Campbell, J.A., Laul, J.C., Nielson, K.K. and Smith, R.D. (1978). Separation and chemical characterization of finely sized fly ash particles. Anal. Chem., 50, 1032-1040.

Cannon, H.L. and Suranson, V.E. (1980). Contribution of major and minor elements to soils and vegetation by coal fired four corners power plant, San Juan Country, New Mexico. Geological Survey Professional Paper, No. 1129: A-I: Bl - B13.

Capp, J.P. (1978). Power plant fly ash utilization of land reclamation in the eastern United States. Pp. 339-353. In: F.W. Schaller and P. Sutton (eds.), Reclamation of Drastically Disturbed Land. ASA, Madison, W.I.

Carlson, C.L. and Adriano, D.C. (1991). Growth and elemental content of two tree species growing on abandoned coal fly ash basins. J. Environ, Qual., 20, 581-587.

Carlson, C.L. and Adriano, D.C. (1993). Environmental impacts of coal combustion residues. J. Environ. Qual., 22:2, 227-247.

Caray, E.E., Gilbert, M., Bache, C.A., Gutenmann, W.H. and Lisk, D.J. (1983). Elemental composition of potted vegetables and millet grown on hard coal bottom ash amended soil. Bull. Environ. Contam. Toxicol., 31, 418-423.

Catalan, R.E., Martinez, A.M. and Aragones, M.D. (1980). Effects of vanadate on the cyclic AMP-protein kinase system in rat liver. Biochem. Biophys. Res. Commun., 96, 673-678.

Carvelli, S., Petruzzelli, G. and Perna, A. (1986). Fly ashes as an amendment in cultivated soils: Effect on mineralization and nitrification. Water, Air and Soil Pollut., 22, 331-338.

Chandrasekar, V. and Narayanswami, R. (1993). Effect of potash and amendments of the tikka leaf spot disease incidence of groundnut. J. Porass. Res., 9:4, 370-374.

Chaney, R.L., Honik, S.B. and Simon, P.W. (1976). Heavy metal relationship during land utilization of sewage sludges in the Northeast. In : Proc. Cornell Agri. Waste Manag. Conf.

Chang, A.C. Page, A.L. and Warneke, *J. E.* (1987). Long term sludge applications on Cadmium and Zinc accumulation in swiss chard and radish. J. Environ. Qual., 16, 217- 221.

Chang, A.C., L.J. Lund, A.L., Page and J.E. Warneke., (1977). Physical properties of fly ash amended soils. J. Environ. Qual., 6, 267-270.

Chang, A.C., Page, A.L., Lund, L.J., Warneka, J.E. and Nelson, C.O. (1989). Municipal sludges and utility ashes in California and their effects on soils In: B. Bar-Yosef *et al* (eds.) Inorganic contaminants in the Vadose Zone. Ecological Studies, Vol. 74, Springer-Verlag, Berlin, pp. 125-129.

Chapman, H.D. (1965). Diagnostic criteria for plants and soils. Abilen, Tex, Quality Printing Co.

Cherenkova, E. (1977). Power station fly ash applied to the soil and its effect on plants. Rostlinna vyroba, 23: 4, 429-438.

Chermisri, C., Watanabe, H., Attajarusit, S., Tuntiwarawit, J. and Kacwroj, S. (1995). Effect of boron sources on garlic *(Allium sativuam* L.) productivity. Biol. Pert. Soils, 2O:2, 125-129.

Cope, F. (1962). The development of a soil from an industrial waste ash. In : Soc. Soil Sci. Trans. Comm. IV, Palmerstown, New Zealand, pp. 859-863.

Coreil, C.B., Boquet, D.J. and Breitenbeck, G.A. (1997). Soil applications of papermill and municipal, biosolids for cotton production. In: Proceedings Beltwide Cotton Conference, New Orleans, A, USA, Volume 1, 652-655.

Croxdale, J. *L* (2000). Stomatal patterning in angiosperms. American J. Bol., 87, 1069- 1080.

Dabas, S. and Singh, R. P. (1995). Differential effect of lead on nitrate reducates activity and organic nitrogen content in mung bean (Var. P 105) Seedlings. Indian J. Plant Physiol., 38, 155-157.

Elseewi, A.A. Straughan, I.R. and Page. A.L. (1980). Sequential cropping of fly ash amended soils: Effects on soil chemical properties and yield and elemental composition of plants. Sci. Total Environ., 15, 247 *259*.

Elseewi, A.A., Bingham, F.T. and Page A.L. (1978). Availability of sulfur in fly ash to plants. J. Environ. Qual, 7:1, 1978.

Elseewi, A.A., Bingham, F.T. and Page, A.L. (1978). Growth and mineral Composition of lettuce and Swiss chard grown on fly ash amended soils. In : D.C. Adriano and I.L. Brisbin (eds.) Environmental Chemistry and Cycling Processes. CONF 760429. US Dep. Commerce. Springfield, pp. 568-581.

Elseewi, A.A., Grimm, S.R., Page, A.L. and Straughan, I.R. (1981). Boron enrichment of plants and soils treated with coal ash. J. Plant Nutrition, 3:1-4, 409-427.

Fail, J.L. (1987). Growth response of two grasses and a legume on coal fly ash amended strip mine spoils. Plant Soil, 1O1, 149-150.

Fail, J.L. and Wockhok, Z.S. (1977). Soyabean growth on fly ash amended strip mine spoils. Plant Soil, 48, 473-484.

Fine, L.O. and Jensen, W.P. (1981). Phosphate in waters. I. Reduction using northern lignite fly ash. Water Resour. Bull., 17, 895-897.

Fletcher, R.A., Adedipe, N.D. and Ormord, D.P. (1972). Abscissic acid protects bean leaves from ozone induced plytotoxicity. Can. J. Bot., 50, 2389-2391.

Fluckiger, W., Dertli, J.J. and Fliickiger-Kelker, H. (1979). Relationship between stomatal diffusive resistance and various applied particle sizes on leaf surfaces. *Z. Pflanzen physiol.,* 91, 173-175.

Fox, J.E.D., Doronila, A.I. and Ownes, B.K. (1990). Growth of three subterranean clover *(Trifolium subterraneum)* cultivars on coal mine interburden materials amended with fly ash. Mulga Res. Cent. J., 10, 45-49.

Francis, C.W., Davis, E.C. and Goyert, J.C. (1985). Plant uptake of trace elements from coal gasification ashes. J. Environ. Qual., 14, 561-569.

Gangloff, W.J., Ghodrati, M., Sims, J.T. and Vasilas, B.L. (2000). Impact of fly ash amendments and incorporation method of hydraulic properties of a sandy soil. Water, Air and Soil Pollut., 119 : 1-4, 231-245.

Ganju and Puri (1959). In: The Wealth of India, Raw materials (1962), Vol. VI, Publication and Information Directorate, CSIR, New Delhi.

Ghodrati, M., Sims, J.T., Vasilas, B.L. and Hendricks, S.E. (1995). Enhancing the benefits of fly ash as a soil amendment by pre leaching. Soil Sci. 159:4, 244-252.

Ghose, M.K. and Banerjee, S.K. (1995). Status of air pollution caused by Coal Washery Projects in India. Environ. Monit. Assess., 38, 97-105.

Ghouse, A.K.M. and Yunus, M. (1972) Preparation of epidermal peels from leaves of gymonsperms by treatment with hot 60% HNO.,. Stain TechnoL, 47, 322-324.

Ghuman, G.S., Menon, M.P., Chandra, K., James, J., Adriano, D.C. and Saiwan, K.S. (1994). Uptake of multi elements by corn from fly ash compost amended soil. Water Air Soil Pollut., 72, 285-295.

Gibon, Y., Sulpice, R and Larher, F. (2000). Proline accumulation in Canola leaf discs subjected to osmotic stress is related to the loss of Chlorophylls and to the decrease of mitochondrial activity. Physiol. Plant., 110, 469-476.

Giedrojc, B.; Fatyga, J. and Hryncewics, Z. (1980). The effect of fertilization with ashes from black coal burned in electric power station on the properties of sandy soil and crop. Polish J. Soil Sci., 13:2, 163-171.

Gilliam, T.M., Canon, R.M., Egan, B.Z., Kelmers, A.D., Seeky, F.G. and Watson, J.S. (1982). Economic metal recovery from fly ash. Resour. Conserv., 9, 155-168.

Gouia, H., Ghorbal, M. H and Meyer, C. (2000). Effects of Cadmium on activity of nitrate reductase and on other enzymes of the nitrate assimilation pathway in bean. Plants Physiol. Biochem., 38, 629-638.

Gray, D.H. and Lin, Y.K. (1972). Engineering properties of compact fly ash. J. Soil Mechan. Found Div., Am. Soc. Cir. Eng., 98, 361-380.

Griffith, E.D. and Clarke, A.W. (1979). World coal production. Scient. Amer., 240, 38-47.

Grover, H.L., Nair, T.V.R. and Aborl, Y.P. (1978). Nitrogen metabolism of upper three leaf blades of wheat at different soil nutrient levels. NR activity and content of various nitrogen constituents. Plant Physiol., 42, 287-292.

Gupta, A.P., Antil, R.S., Narwal, R.P. and Singh, A. (1996). Utilization of applied phosphorus by wheat as influenced by fly ash. Ann. Biol. Ludhiana, 12:1, 19-24.

Gupta, G.K. and Singh, D. (1996). Effectiveness of water absorbents, FYM application and sowing direction in downy mildew reduction in pearl millet under arid environment. Ind. Phytopath., 49:1, 41-47.

Gupta, U.C. (1993). Deficiency and toxicity levels of boron in crops. In : Boron and its role in crop production, CRC Press, Boca, Raton FL. Umesh C. Gupta (ed.), pp. 137-145.

Gupta, U.C. and Gupta, S.C. (1998). Trace element toxicity relationships to crop production and livestock and human health: implications for management. Commun, soil sci. plants analys., 29: 11-14, 1491-1522.

Gupta, U.C., Sterling, J.D.E. and Nass, H.G. (1973). Influence of various rates of compost and nitrogen on the boron toxicity symptoms in barley and wheat. Con. J. Plant, Sci., 53, 451-456.

Gutenmann, W.H. and Lisk, D.J. (1996). Increasing Selenium in field grown onions by planting in peat moss pots containing coal fly ash. Chemosphere, 32:9, 1851-1853.

Gutenmann, W.H., Elfving, D.C., Valentino, D.I. and Lisk, D.J. (1981). Trace element absorption on soil amended with soft coal fly ash. Bio Cycle, 22.

Hale, J.C., Ormrod, D.P., Laffey, P.J. and Alien, O.B. (1985). Effect of nickel and copper on tomato in sand culture. Environ. Pollut., 39, 53-69.

Hammer meister A.M., Naeth, M.A. and Chanasyk, D.S. (1998). Implications of fly ash application to soil for plant growth and feed quality. Environ. Technol., 19:2, 143-152.

Hampp, R., Beulich, K. and Ziegler, H. (1976). Effects of Zinc and Cadmium on the photosynthetic CO_2-fixation and Hill activity of isolated spinach chlorplasts. *Z. Pflanzenphysiol,* 77, 336-334.

Hanson, R.L., T.R. Henderson, C.H. Hobbs. C.R. Clark, R.L. Carpenter. J.S. Dutcher, T.M. Harvey and Hunt D.F., (1983). Detection of nitroaromatic Compounds on Coal Combustion particles. J. Toxicol. Environ. Health, 11, 971-980.

Haq, S., Singh, R.P. and Saxena, S.K. (1995). Control of the root knot nematode *Melnidogyne incognita* on tomato with soft coal fly ash. Inter. Nematol. Net. News, 2:3, 4-6.

Harris, W.R., J.F. Remsen, E.K. Chess and D.W. Later, (1987). Corrleation of nitro aromatic compounds with mutagenic activity of Coal fly ash. Environ Mutagen, 6, 131-144.

Harrison, F.L., D.J. Bishop and Mallon B.J., (1985). Comparison of organic, Combustion products in fly ash Collected by a venturiwet scrubber and an electrostatic precipiattor at a Coal fired power station. Environ. Sci. Technol., 19, 186-193.

Heath, R.H. and Packer, L. (1968). Photoperoxidation in isolated Chloroplasts. I. Kinetics and stoichiometry of fatty acid peroxidation. Arch. Biochem. Biophys., 125, 189-198.

Helder, T., Stutterheim, E., Olte, K. (1982). The toxicity and toxic potential of fly ash from municipal incinerators assessed by means of a fish early life stage test. Chemosphere. 11, 965-72.

Hemandez, L.E., Garate, A. and Carpena -Ruiz, R. (1997). Effects of Cadmium on the uptake, distribution and assimilation of nitrate in Pisum sativum. Plant soil, 189, 97-106.

Henning, S.J., Taber, H.G. and Hansen, R.J. (1988). Fly ash liming effect on five crops grown on sandy irrigated soil. Proc. Ann. Madison waste Conf., 11, 323-332.

Hermann, J. (1996). fly ash as boron source to plants. VIII Symposium Microelements in Agriculture. *Zeszyty Problemowe Postepow Nank Rolniczych,* 434: I, 193-198.

Higgins, S.P. and Fey, M.V. (1993). Selenium in some grass postures of Natal and fertilizer induced Se uptake by rye grass *(Lolium multiforum).* South African J. Plant and Soil, 10:4, 186-192.

Hill, M.J. and Lamp, C.A. (1980), Use of pulverised fuel ash from Victorian brown coal as a source of nutrients for a pasture species. Aust. J. Exp. Agric. Anim. Husb., 20, 377-384.

Hiscox, J.D. and Israelstain (1979). A method for extraction of chlorophyll from leaf tissue without maceration. Can.J. Bot., 57, 1332-1334.

Hodgson, D.R. and Buckley, G.P. (1975). A practical approach towards the establishment of trees and shrubs on pulverised fuel ash. In : *M.J.* Chadwick and G.T. Goodman (eds.). The Ecology of

Resource Degradation and Renewal. Blackwell Scientific, Oxford, pp. 305-329.

Hodgson, D.R. and Holliday, R. (1966). The agronomic properties of pulverised fuel ash. Chem. Ind., 20, 785-790.

Hodgson, L., Dyer, D. and D.A. Brown : (1982). Neutralization and dissolution of high Calcium fly ash. J. Environ. Qual., 11, 93-98.

Hortensteiner, S. (1999). Chlorophyll breakdown in higer plants and algae, cell Mol. Life sci., 56, 330-347.

Horton, J.H. and McMinn, J.W. (1977). Ash basin reclamation with forest trees. Savannah River Lab. Doc. DP - 1477, Aitken, S.C.

Hudson, C. and Headley, Q. (1990). Composts as soil ameliorants. In Proceedings Annual conference Barbados Society of Technologist in agriculture, 35-37.

Iqbal, M., Jacob, M., Mehta, S. and Zafar, M. (2000). Plant responses to particulate matters suspended in the atmosphere. In: Environmental Hazards-Plants and People. Iqbal, M., Srivastava, P.S. and Siddiqi, T.O. (eds.), 80-112.

Jacobs, L.W., Erickson, A.E., Berti, W.R. and Mackellerm, B.M. (1991). Improving crop yield potentials of coarse textured soils with coarse flyash amendments. In: Proc. Ninth Int. Ash use Symposium, Am. Coal Ash Assoc., Washington DC, 3, 59.1-16.

Jain, M. and Gadre, R. P. (1998). Inhibition of chlorophyll synthesis and enzymes of nitrogen assimilation by selenite in excised maize leaf segments during greening. Water Air Soil pollut., 104, 161-166.

Jastrow, J.D., Zimmerman, C.A., Dvorak, A.J. and Hinchman, R.R. (1981). Plant growth and trace element uptake on acidic coal refuse amended with lime of fly ash. J. Environ. Qual., 10:2, 154-160.

Jiang, R.F., Yang, C.G., Su, D.C. and Wong, J.W.C. (1999) Coal fly ash and lime stabilized biosolids as an ameliorant for boron deficient acidis soils. Environ. Technol., 20:6, 645-649.

Jones, L.H. and Lewis, A.V. (1960). Weathering of fly ash. Nature (Lodon), 185, 404-405.

Joshi, U.N., Arora, S.K. and Luthra., Y.P. (1993). SO_2 induced changes in CO_2 fixation and phtosynthetic pigments in *Sorghum bicolor* L. leaves. Ann. Biol, 9, 102-108.

Kaakinen, J.W., Jorden, R.M., Lawasani, M.H. and West, R.E. (1975). Trace element behaviour in coal fired power plant. Environ. Scr. Technol, 9, 862-869.

Kabata-Pendias, A. and Pendias, H. (1985). Trace elements in soil and planes. C R C press, Boca Raton, Florida.

Kaiser, W.M., Weiner, H. and Huber, S.C. (1999). Nitrate reductase in higher Plants : A Case study for transduction of environmental stimuli into Control of Catalytic activity. Physiol. Plant, 105, 385-390.

Kamenova, S., Georgieva, V., Merkchiiska, M., Georgieva, N., Boichmova. M. and Paunova, S. (1995). Direct and residual effect of fly ash from a thermo-electrical station on some biological responses of bean plants grown on different soil types. *Rasteniev dni Nauki,* 32:3, 51-54.

Kastori, K. , Petrovic, M. and Petrovic. N. (1992). Effect of excess lead, Cadmium, copper, and Zinc on Water relations in sunflower..J. Plant Nutr., 15, 2427-2439.

Katzur, J. and Gora, E. (1986). Effects of different application rates of hard coal filters ashes on soil, crop yields and mineral fertilization. *Archiv fur Acker and Pflanzenbau und Bodekunde,* 30:5, 301-309.

Kaweeki, Z., Tomas zewska., Z., Smolarz, K. (ed.), Zmarlicki, K. (1993). The effect of mineral and organic fertilisation and of fly ashes on growth and yield of black currant. In: *Sixth Inter. Symp. Rubus and Ribes, Skiernivwice, Pol and Acta Horticultural,* 352, 61-63.

Keisling, T.C., Thompson, L.F. and Slabaugh, W.R. (1984). Visual symptoms and tissue manganese concentrations associated with manganese toxicity is wheat. Commun. Soil Sci. Plant Anal., 15 537-540.

Kene, D.R., Lanjewar, S.A., Ingole, B.M. and Chaphale, S.D. (1991). Effect of application of fly ash on physico - chemical properties of soils. J. soils crops, 1:1, 11-18.

Keshan, U. and Mukherji, S. (1992). Effect of Cadmium toxicity on chlorophyll Content, Hill activity and chlorophyllase activity in *Vigina radiata* L. leaves. Indian Plant Physiol., 35, 225-230.

Khalil, S. and Shah, H.M, (1979). Effect of soil moisture and pH on root-knot disease of tobacco. Tobacco Research, 5, 129-34.

Khan, M.R. and Ghadirpour, M.H. (1999). Effects of field application of fly ash on the growth and yield of chilli eggplant and tomato in the presence and absence of root knot nematode. *Zeitschrift fur Pflanzenkrankheiten and Pflanzenschtuz,* 106:6, 624-632.

Khan, M.R. and Khan, M.W. (1994). Single and interactive effects of O_3 and SO_2 on tomato. Environ. Expt. Bot., 34, 461-469.

Khan, M.R. and Khan, M.W. (1996). The effect of fly ash on plant growth and yield of tomato. Envir. Pollut.,. 92:2, 105-111.

Khan, M.R.. Khan, M.W. and Singh, K. (1997). Management of root knot disease of tomato by the application of fly ash in soil. Plant Pathol., 46: 1, 33-43.

Khan, S., Begum, T. and Singh, J. (1996). Effect of fly ash on physico-chemical properties and nutrient status of soil. Ind. J. Environ. Hlth., 38, 41-46.

Khandkar, U.R., Gangwar, M.S., Srivastava, P.C. and Singh, M. (1996). Effect of coal fly ash application on the elemental composition and yield of some crops and on the properties of a calcareous soil. Acta Agron. Hungarica, 44:2, 141-151.

Klein, D.H., Andren, A.W., Carter, J.A., Emery, J.F., Feldman, C., Fulkerson, W., Lyon, W.S., Ogle, J.C., Talmi, Y., Van Hook, R.I. and Bolton, N. (1975). Pathways of thirty seven trace elements through coal tired power plants. Environ. Sci. Technol., 9, 973-979.

Klein, M., chang, G., Chung, M. and Tallman, G. (1996). Effects of turgor potentials of epidermal cells neighbouring guard cells on stomatal opening in detached leaf epidermis and intact leaflets of *Vicia faba L.* Plant cell Environ., 19, 1399-1407.

Klepper, L., Flescher, D. and Hageman, R.H. (1971). Generation of reduced nicotinamide adenine dinucleotide for nitrate reduction in green leaves. Plant Physiol., 48, 580-590.

Korcak, R.F. (1984). Effect of coal combustion wastes used as lime substitutes in nutrition of apples in three soils. Plant soil, 85:3, 437-441.

Krajickova, A. and Majstrick, V. (1984). The effect of fly ash particles on the plugging of stomata. Environ. Pollut., 36 A, 83-93.

Kramer, P.J. (1969). Plant, Soil and Water Relationships. A Modern Synthesis : Mc-Grew-Hill. New York.

Kuchanwar, O.D. and Matte, D.B. (1997). Study of graded doses of fly ash and fertilizers on growth and yield of groundnut *(Arachis hypogaea).* J. soils Crops, 7:1, 36-38.

Kuchanwar, O.D., Matte, D.B. and Kene, D.R. (1997). Evaluation of graded doses of fly ash and fertilizers on nutrient content and uptake of groundnut grown on Vertisol. J. soils crops, 7:1, 1 3.

Kukier, U. and Sumner, M.E. (1996). Boron availability to plants from coal combustion by products. Water, Air and Soil Pollut., 87:1-4, 93-110.

Kukier, U., Sumner, M.E. and Miller, W.P. (1994). Boron release from fly ash and its uptake by corn. J. Environ. Qual., 23:3, 596-603.

Kuziemska, I., Quant, B. and Sulek, Z. (1989). Effects of industrial water of fly ash storage from a storage yard on organisms in water of the Gdansk coastal region (Northern Poland). Acta Hydrobiological. 30: 3-4, 317-328.

Lagerwerff, J.V. (1971). Uptake of Cadmium, lead, and zinc by radish from soil and air. soil sci., 111, 129-133.

Lal, A., Ku, M.S.B. and Edwards, G.E. (1996). Analysis of inhibition of photosynthesis due to water stress in the *C2* species *Hordeum vulgare* and *Vicia faba.* Electron transport, CO2 fixation and Carboxylation Capacity. Photosyn. Res., 49, 57-69.

Lalk, J.K., Mishra, B., Sarkar, A.K. and Lal, S. (1996). Effect of fly ash on growth and nutrition of soyabean. J. Ind. Soc. Soil Sci., 44 : 310-313.

Lamoreaux, R.J. and Chaney, W.R. (1978). The effect of cadmium on net Photoyrithesis, transpiration, and dark respiration of excised silver maple leaves. Physiol. Plant, 43,231-236.

Lebedev, N.N., Siffel, P. and Kransnovsky, A.A. (1992). Free chlorophylls in green plant cells. Res. Photosyn., 3, 127-130.

Ledig, F.T. (1974). Concepts of growth analysis : In : Proceedings of the 3rd North American Forest Biology Workshop. (Eds : C.P.P. Ried and G.A. Fechmer). Colorado state university. Fort Collins. pp. 166-182.

Lee Jr. R.E., Gist, M.L., Reley, A.E. and Macleod, K.E. (1975). Concentration and size of trace metal emission from a power plant, a steel plant and a cotton gin. Environ. Sci. Technol., 9, 643-647.

Lindon, F.C. and Henriques, F.S. (1998). Role of rice shoot Vacuoles in copper toxicity regulation. Environ. Exp. Bot., 39, 197-202.

Linton, R.W., Loh, A., Natusch, D.F.S., Evans, G.A. and Williams, P. (1976). Surface predominance of trace elements in air borne particles. Science, 19, 852-854.

Low, G.K.C., G.E. Bately, R.O. Lingard and A.M. Dunfield, (1986). Determination of polycyclic aromatic hydorcarbons in coal fly ash using gas chromatography/negative ion chemical iomzation mass spectrometry. Bioed. Environ. Mass. Spect., 13, 95-104.

Macara, I.G. (1980). Vanadium-an element in search of a role. Trends Biochem Sci., April, pp. 92-94.

Machado, M.O., Dias, A.D., Turatti, A.L. and Zonta, E.P. (1984). Effect of coal ash on irrigated rice and on some chemical properties of the soils' *Anais da 13a Reunia da cultura do Arroz Irrigado,* 202-210.

Maclachlan, S. and Zalik, S. (1963). Plastid structure, chlorophyll concentration and free ammo acid composition of a chlorophyll mutant of barley. Can. J. Bot., 41, 1053-1062.

Makino, A., Mae, T. and Ohira, K. (1983). Photosynthesis and ribulose 1, 5 diphosphate carboxylase in rice leaves. Plant Physiol, 73, 1002-1003.

Maldonado, J.M., Barro, F. and Gonzaliz-Fontes, A. (1996). Nitrate assimilation and protein accumulation. Plant Physiol. Biochem., 34, 721-726.

Malewar, G.U., Adsul, P.B. and Ismail, S. (1998). Effect of different combinations of fly ash and soil on growth attributes of forest and dry land fruit crops. Ind. J. Forest., 21:2, 124-127.

Malewar, G.U., Adsul, P.B. and Ismail, S. (1999). Growth parameters in tomato and spinach as influenced by fly ash, soil and their combinations. J. soils and crops, 9:1, 30-33.

Malewar, G.U., Adsul, P.B. and Ismail, S. (1999). Impact of different levels of fly ash on growth attributes and dry matter yield of various crops. J. Maharashtra agric. Univ., 24:2, 220-221.

Mankovska, B. (1992). Chemical composition of solid particles on vegetative surface in slovak forests. *Ekologia,* 11:2, 205-214.

Mankovska, B. (1997). Variations in sulphur and nitrogen foliar concentration of deciduous and coniferous vegetation in Slovakia. Water, Air, Soil Pollut., 96:1, 329-345.

Manz, O.E. (1979). Economics of reuse of power plant ash. In : M.K. Wali (ed.) Ecology and coal Resource Development, Pergamon Press, New York. 1, 339-342.

Mark, L.T. (1963). Temperature inhibition of carotene synthesis in tomato. *Bet. Guz.,* 124, 180-185.

Martens, D.C. (1971). Availability of plant nutrients in fly ash. Compost. Sci., 12, 15-19.

Martens, D.C. and Beahm, B.R. (1976). Growth of plants in fly ash amended soils p. 657-664. In : J.H. Faber *et al.,* (eds). Proc. 4th Int. Ash Utilization Symposium, St. Louis, MO. 24-25 Mar. 1976. MERC SP-76/4. ERDA Morgantown Energy Res. Center, Morgan -town, WV.

Mc Leod, K.W. and Ciravolo, T.G. (1997). Differential sensitivity of Nyssa *aquatica* and *Taxodium distichum* seedlings grown in fly ash amended sand. Wetlands, 17:2, 330-335.

Mc Minn, J.W., Berry, L.E. and Horton, J.H. (1982). Ash basin reclamation with commercial forest species. Reclam. Reveg. Res., 1, 359-365.

Mc Murphy, L.M., Birdar, D.P., Tacts, C. and Rayburn, A.L. (1996). Differential effects of weathered coal fly ash and fly ash leachete on the maize genome. Arch. Environ. Contam. Toxicol., 31:2, 166-169.

Menon, M.P., Sajwan, K.S., Ghuman, G.S., James, J., Chandra, K. (1993). Ash amended compost as a manure for agricultural crops. J. Environ. Sci. Health, 28A:9 2167-2182.

Menzies, N.W. and Aitken, R.L. (1996). Evaluation of fly ash as a component of potting substrates. *Scientia Horticultural,* 67: 1-2, 87-99.

Merakchiiska, M., Kamenova, Y.S. and Georgieva, V. (1995). Changes in the pigment content, fluorescence and photosynthetic activity of the leaves of bean plants *(Phaseolus vulgaris* L.) as a result of the direct and residual effects of fly ash added to soils. *Rasteniey" dni Nauki,* 32: 3, 55-59.

Minguez-Mosguera, M.I. and Gullardo Guerrero, L. (1996). Role of chlorophyllase in chlorophyll matabolism in olives Cv, Gordal. Phytochemistry, 41, 691-697.

Mishra, A.K., Bhowmik, A.K. and Banerjee S.K. (1996). Effect of mulches on growth of tree species on fly ash. Env. Ecol., 14:2, 411-414.

Mishra, A.K., Williams, A.J. and Banerjee, S.K. (1995). Mycorrhizal development and plant growth in amended fly ash of thermal power plants. Adv. Fores. Res. Ind., 12, 213-227.

Mishra, L.C. (1982). Effect of environmental pollution on the morphology and leaf epidermis of *Commelina bengalensis* Linn. Environ. Pollut., 28:4, 281-284.

Mishra, L.C. and Shukla, K.N. (1986). Effects of fly ash deposition on growth, metabolism and dry matter production of maize and soyabean. Environ. Pollut., 42 A, 1-13.

Mishra, P.K., Das, R.K. and Mehta, U.C. (1999). Mitotic study of radish grown on flyash amended soil. Cruciferae Newsletter, 21,49-50.

Mishra, S.N. and Srivastava, H.S. (1983). Increase in nitrate reductase activity, organic nitrogen and protein content of maize leaves supplied with nitrate and ammonium. *Z. Pflanzenphysiol,* 113, 91-93.

Mishra, S.N., Bhutani, S. and Singh D.B. (1994). Influence of nitrate supply on Cadmium toxicity in *Brassica juncea* during early seedling growth. Indian J. Plant Physiol., 37, 12-16.

Moffat, A.J. and Shaw, P.J.A (1993) Establishment of trees on mixture of pulverized fuel ash and gypsum. Land Degrad. Rehab., 4:2, 87-97.

Moliner, A.M. and Street, J.J. (1982). Effect of fly ash and lime on growth and composition of corn *(Zea mays)* on acid sandy soils. Proc. Soil Crop Sci. Soc. Florida, 41, 217-220-

Mooree, R.D. Wheeler, D.E. and Wilber, K.R. (1978). A comparison of drift transport predictins for various types of cooling systems. In : Cooling Tower Environment-1978. A Symp. on Environmental Effects of Cooling Tower Emissions, Maryland Dept. Nat., Resour. Power Plant Siting Program and Water Resour. Res. Ctr. Univ. of Maryland, College Park. 231-243.

Morghan, J.T. (1979). Manganese, toxicity in flax growing on certain calcareous soil low in available iron. Soil Sci. Soc. Am. J., 43, 1177-1180.

Moya, J.L., Ros, R. and Picago, I. (1995). Heavy metal harmone interactions in rice plants : Effects on growth, net photosynthesis. and carbohydrate distribution. J. Plant Growth Regul., 14, 61-67.

Mudd, J.B. (1975). Peroxyacyl nitrates. In : J.B. Mudd and T.T. Kazlowski (eds.). Responses of plants to air pollution, Academic Press, New York, pp. 97-119.

Mulchi, C.L. and Armbruster, J.A. (1981). Response of corn and soyabeans to simulated saline aerosol drift from brackwish water cooling towers. J. Environ. Qual., 10, 541-547.

Mulford, F.R. and Martens, D.C. (1970). Responses of alfalfa to boron in fly ash. Soil Sci. Am. Proc., 35, 296-300.

Murray, F. (1984). The accumulation by plants of emissions from a coal tired power plant. Atmos. Environ., 18, 1705-1709.

Murtha, M.J., Burnet, G. and Harnby, N. (1983). Power plant fly ash-disposal and utilization. Environ. Progr., 2, 193-198.

Nafees, G. (1995). Plant growth and nodulation in *Vicio fobo* L. under sodium chloride stress. J. Indian Bot. Soc., 74, 341-342.

Nag, P., Paul, A.K. and Mukherjee, S. (1981). Heavy metal effects in plant tissues involving chlorophyll, chlorophyllase, Hill reaction activity and gel electrophoretic patterns of soluble proteins. Indian J. Exp. Biol., 19,702-706.

Nass, M.M., Lexmond, T.M., Van Beusichem, M.L. and Jansen Turkovicova, M. (1993). Long term supply and uptake by plants of elements from coal FA. Commun. Soil Sci. Plant Analy., 24:9-10, 899-913.

Natusch, D.F.S., Wallace, J.R. and Evans, C.A. (1975). Toxic trace elements : Preferential concentration in response to particles. Science, 183, 202-204.

Nilsen, E.T. and Orcutt, D.M. (1996). The Physiology of Plants under stress. John Wiley and Sons., Inc., New York.

Nwosu, J.U., Harding, A.K. and Linder, G (1995). Cadmium and lead uptake by edible crops grown in a silt loam soil. Bull. Environ. Contam. Toxicol., 54, 570-578.

Oblisami, G., Pathmanabhan. G. and Pathmanabhan, C. (1978). Effects of particulate pollutant from cement kiln on cotton plants. Indian J. Air Pollut. Cont, 1, 91-94.

Osmond, C.B., Austin, M.P., Berry, J.A. Billings, W.D., Boyer, J.S., Dacey, J.W.H., Nobel, P.S., Smith, S.D., and Winner W.E. (1987). Strees physiology and the distributions of plants. Bio Sci., 37, 38-48.

Oswal, M.C., Singh, C.B. and Grewal, K.S. (1999). Potential of fly ash as a soil conditioner for desert ecosystem. In : Faroda, A.S Joshi, N.L., Kathju, S. and Amal, K. (eds.). Recent advances in management of arid ecosystem. Proceedings of a symposium held in India, 1997, 469-474.

Padmakaran, P., Raju, C.B. and Rao, A.S. (1994). Characterization of fly ash to assess its sensibility for agricultural purposes. Clay Res., 13:1-2, 30-37.

Page, A.L. and Chang, A.C. (1979). Contamination of soil and vegetation by atmospheric deposition of trace elements. Phyto pathol., 69, 1007-1011.

Page, A.L., A.A. Elseewi and L.R. Straughan, (1978). Physical and chemical properties of fly ash from coal, fired power plants with reference to environmental impacts. Residuce. Rev., 71, 83-120.

Pandey, V., Mishra, J., Singh, S.N., Yunus, M., Ahmad, K.J. and Singh N. (1994). Growth response of *Helionthus annuus* L. grown on fly ash amended soil, J. Environ. Biol., 15:2, 117-125.

Parekh, D., Puranik, R.M. and Srivastava, H.S. (1990). Inhibition of chlorophyll biosynthesis by Cadmium in greening maize led segments. *Biochem. Physiol. Pflanzen.,* 186, 239 242.

Pathak, D., Gupta, B.K. and Bhattacharya, A.K. (1989). Accumulation and movement of Ca and Mg in fly ash amended soil. Environ. Pollut., 62, 57-61.

Patil, C.V., Math, K.K. Bulbule, A.V. Prakash, S.S. and Yeledhalli, N.A. (1996). Effects of fly ash on Soil crust strenght and crop yield. J. Maharashtra agric. Univ., 21:1, 9-11.

Patil, C.V., Math, K.K., Bulbule, A.V., Prakash, S.S. and Yeledhalli, N.A. (1996). Preliminary investigations on the growth of *Albizia lebbeck* Bentn. on blended iron ore overburden. Ind. J. Soil Conser., 24:1, 44-48.

Patterson, G.D. Jr. (1958) Sulphur in chlorometric determination of non metals. In : D.F. Blotzl (ed.), Inter. Sci. Pub. Inc., New York.

Pawar, K. and Dubey, P.S. (1988). Germination behaviour of some important crop species in flyash incorporated soils. In : Verma, G.P. (eds.), Advancement of crops and monitoring of environment : Progress in Ecology, 10, 295-305.

Pawar, K., Trivedi, L. and Dubey, P.S. (1982). Comparitive effects of cement, coal dust and fly ash on *H. abelmoschus.* Int. J. Environ. Stud.

Pawlik, B., Skowronski, T., Ramazanow, Z., Gardestrom, P. and Samuelsson, G. (1993). pH dependent Cadmium transport inhibits photosytbesis in the cyanobacterium *Synechocystis aquatilis.* Environ. Exp. Bot., 33, 331-337.

Perkins, P.V. and Vann, A.R. (1995). The early performance of deciduous trees grown in acidic minesoil amelioreted with pulverised fuel ash. Land Degrad. Rehab., 6:1, 57-67.

Perronnet, K., Schwartz, C., Gerard, E. and Morel, J.L. (2000). Availability of Cadmium and Zinc accumulated in the leaves of *Thlaspi caerulesans* incorporated into soil. Plant soil, 227, 257-263.

Petrikova, V., Vana, J. and Ustjak, S. (1996). Production and utilization of industrial energy crops and reclaimed soils. *Metodiky- pro-Zemedelskou-Praxi,* 17, 24pp.

Petruzzelli, G., Lubrano, L. and Cervelli, S. (1987). Heavy metal uptake by wheat seedlings grown in fly ash amended soils. Water, Air and Soil Pollut., 32: 3-4, 389-395.

Phung, H.T., Lund, L.J. and Page, A.L. (1978). Potential use of fly ash as a liming material. In : D.C. Adriano and I.L. Brisken (eds.). Environ. Chem. Cycl. Processes. CONF 760429. US Dep. Commerce. Springfield, Va. pp. 504-515.

Pichtel, J. Kuroiwa, K. and Sawyerr, H.T. (2000). Distribution of Pb, Cd and Ba in soils and plants of two contaminated sites. Environ. Pollut., 110, 171-178.

Pietz, R.I., Carlson Jr., C.R., Peterson, J.R., Zenz, D.R. and Lue-Hing, C. (1989a). Application at sludge and other amendments 10 coal refuse material: I. Effect on chemical composition. J. Environ. Qual., 18, 164-168.

Pietz, R.I., Carlson Jr., C.R., Peterson, J.R., Zenz, D.R. and Lue-Hing, C. (1989b). Application of sewage sludge and other amendments to coal refuse material : II. Effects on revegetation. J. Environ. Qual., 18, 169-173.

Pietz, R.I., Carlson Jr., C.R., Zenz, D.R. and Lue-Hing, C. (1989c). Application of sewage sludge and other amendments to coal refuse material. III. Effects on percolate waste composition, 18, 174-179.

Plank, C.O. and Martens, D.C. (1974). Boron availability as influenced by application of fly ash to soil. Soil Sci. Soc. Am. Proc., 38, 974-977.

Plank, C.O., Martens, D.C. and Hallock, D.L. (1975). Effect of soil application on fly ash on chemical composition and yeild of corn *(Zea mays)* and on chemical composition of displaced soil solutions. Plant Soil, 42, 465-476.

Podgiaskii, M. and Maly, V. (1986). Yields of farm crops after reclamation of fly ash from power station sludge beds. *Scientia Agricultrae Bohemoslouaca,* 18:3, 169-180.

Poison, d.E. and Adams, M.W. (1970). Agron. J., 62, 55.

Polyak, K., Bodog, I. and Hlavay, J. (1994). Determination of chemical species of selected trace element in fly ash. Talanta-Oxford, 41:7, 1151-1159.

Prasad, B.K., Banerjee, S.K. and Roy, H. (2000). Effect of fly ash amendment on germination behaviour and seedling survival on certain cultivated legumes. Environ. Ecol., 18:1, 210-216.

Prasad, D.D.K and Prasad, A.R.K. (1987). Altered 5. amino.-levulinic acid metabolism by lead and mercury in germination seedlings of bajra *(Pennisetum typhoideum).* J. Plant Physiol., 127, 241-249.

Prasad, D.D.K and Prasad, A.R.K. (1987). Effect of lead and Mercury on chlorophyll synthesis in mung bean seedlings. Phytochemistry, 26, 881-883.

Prausse, A. (1991). Yield reduction in large scale agricultural enterprise through long term industrial emissions of boron, sulphur and inert carbon In: *Unweltaspekte der Tierproduction. Vortrage Zum Generalthema des 103. VD LUFA-Kongresses vom* 16-21, 9, 719-724.

Purves, D. and Mac Kenzie, E.J. (1973). Effects of applications of municipal compost on uptake of copper, Zinc and boron by garden vegetables. Plant Soil, 39, 361-371.

Raizada, A., Jayaram, N.S. and Rao, M.S.R.M. (1996). Preliminary investigations on the growth of *Albizia lebbeck* Bentn. on blended iron ore overburden. Ind. J. Soil Conser., 24:1, 44-48.

Raman, S. Patel, A.M., Shah, B.G. and Kaswala., R.R. (1996). Feasibility of some industrial cuastes for soil improvement and crop production. J. Ind. Soc. Soil Sci., 44:1, 147-150.

Ramsden, A.R. and Shibaoka, M. (1982). Characterization and analysis of individual fly ash particles from coal-fired power stations by a combination of microscopy, electron microscopy and quantitative electron microprobe analysis. Atmos. Environ., 16, 2191-2206.

Rangaswami, G. and Bagyaraj, D.J. (1993). Agricultural Microbiology, 2nd Ed. Prentice Hall of India Pvt. Ltd., New Delhi.

Rauser, W.E. and Dumbroff, E.B. (1981). Effects of excess cobalt, nickel and Zinc on the water relations of *Phaseolus vulgaris.* Environ. Exp. bot., 21, 249-255.

Reiberk, H. and Kohrnak, L.V. (1984). Use of coal ash for Casuarina forest biomass production. Soil Crop Sci. Soc. Florida Proc., 43, 76-80.

Rengifo, J.C., Ramirez, G.L. and Bruzon, S.F.C. (1996). Importance of filter press cake as a substrate for vegetable seedling production. *Acta Agronomica,* 46:1-4, 37-43.

Ribaric, C., Batic, F., Dejanovic, B. and Simoncic, P. (1999). Change in the condition of Norway spruce forests after installation of desulphurization devices at the Sostanj Thermal Power Plant. J. Forest Sci., 45:5,217-222.

Rippon, J.E. and Wood, M.J. (1975). Mocrobiological aspects of pulverized fuel ash. In : M.J. Chadwick and G.T. Goodman (eds.), The ecology of Resource Degradation and Renewal. John Wiley and son. New York, pp, 331-349.

Robinson, B.H., Leblanc, M.,Petit, D., Brooks, R.R., Kirkman, J.H. and Gregg, P.E.H. (1987). The potential of *Thlaspi caerulesens* for phytoremediation of contaminated soils. Plnt. soil, 203, 47-56.

Robinson, M.F., Heath, J. and Mansfield, T.A. (1988). Disturbances in stomatal behaviour caused by air pollutants. J. Exp. Bot., 49,461 -469.

Rohrman, F.A. (1971). Analyzing the effect of fly ash on water pollution. Power; 115, 76-77.

Romer, W., Jank, H.J. and Demele, H. (1992). Heavy metal contents of grassland soils and plants near a coal fired power plant. Agribio. Res., 45:2, 137-146.

Rouze, P. and Caboche, M. (1992). Nitrate reduction in higher plants: Molecular approaches to function and regulation. In: Inducible Plant proteins, (Ed. J.L. Wray). Cambridge University press, pp. 45-77.

Saba, Pande, D., Iqbal, M. and Srivastava, P.S. (2000). Effect of $ZnSO_4$ and $CuSO_4$ on lepidine content in *Lepidine sativum* L. Biol. Plant, 43, 253-256.

Sahi, A.N. and Singh, R.N. (1996). Fly ash induced chromosomal aberrations in *Allium cepa,* Cytobios, 86: 3-4, 23-28.

Sajwan, K.S., Ornes, W.H. and Youngblood, T. (1995). The effect of fly ash/sewage sludge mixtures and application rates on biomass production. J. Environ. Sci. Health, 30A : 6, 1327-1337

Sajwan, K.S., Ornes, W.H. and Youngblood, T.V. (1996). Growth and elemental composition of Sorghum Sudangrass grown on fly ash/ organic waste amended soils. J. Environ. Sci. Health, 31A : 7, 1729-1739.

Sale, L.Y., Naeth, M.A. and Chanasyk, D.S. (1996). Growth response of barely on unweathered fly ash amended soil. J. Environ. Qual., 25: 4, 684-691.

Salisbury, E.J. (1927). On the causes and ecological significance of stomatal frequency with special reference to the wood land flora. Phil. Trace R. Soc., 216, 1-65.

Salter, P.J. and Williams, J. B. (1967). Effect of pulverized fuel ash on the moisture characteristics of soils. Nature (London), 213: 1157-1158.

Salter, P.J., Webb, D.S. and Williams, J.B. (1971). Effect of pulverised fuel ash on the moisture characteristics of coarse textured soil and on crop yields, J. Agric. Sci. Comb, 77: 53-60.

Sandmann, G. and Roger, P. (1980). Copper-mediated lipid peroxidation processes in photosynthetic membranes. Plant Physiol., 66, 797-800.

Satayanarayana, G., Bhatnagar, A.R. and Acharya, U.H. (1988). Effect of fly ash pollution on *Datura innoxia.* Environ, and Ecol., 6:1, 92-95.

Satyanarayana, G. and Pushpalatha, K. (1991). Fly ash induced foliar injury and proline metabolism in plants growing near power house. Adv. Plant Sci., 4:2, 298-303.

Sawyer, T., Bandiera S., Safe S., Hutzinger O and Olie, K. (1983). Bioanalysis of polychlorinated dibenzofuran and dibenzo-p-dioxin mixture in fly ash. Chemosphere. 12, 529-36.

Saxena, M., Asokan, P., Aparna C., Saxena M. and Chauhan. (1998). Effect of fly ash on clayey soil. Clay Research, 17:2, 109-114.

Scanlon, D.H. and Duggan, J.C. (1979). Growth and element uptake of woody plants on fly ash. Environ. Sci. Technol., 13, 311-315.

Schinley, W.W. (1979). Waste products : A new trend in reclamation. In : M.K. Wali (ed.) Ecology and coal resource development Vol. 2, Pergamon Press, New York, pp. 967-972.

Schnappinger, Jr. M.G., Martens, D.C. and Plank, C.O. (1975) Zinc availability as influenced *by* application of fly ash to soil. Environ. Sci. Technol., 9:258-261.

Scotti, I.A., Lombi, E., Carini, F. and Silva, S. (1996). Effect of fly ash application on the yield production and chemical composition of chicory. Agrochemica, 40:5-6, 293-303.

Scotti, I.A., Silva, S. and Botteschi, G. (1999). Effects of fly ash on The availability of Zn, Cu, Ni and Cd to chicory. Agri. Ecosys. Environ., 72:2, 159-163.

Sess, J.E. (1958). Botanical Microtechnique, Oxford and IBH Publ. Co., Calcutta.

Setia, R.C., Bala, R. Setia, N. and Malik, C.P. (1993). Photosynthetic characteristics of heavy metal treated wheat *(Triticum aestivum* L.) plants. J. Plant Sci. Res., 9, 47-49.

Shane, B.S., Littman, C.B., Essick, L.A., Gutenmann, W.H. Doss, G.J. and Lisk, D.J. (1988). Uptake of selenium and mutagens by vegetables grown in fly ash containing greenhouse media. J. Agric. Food Chem., 36, 328-333.

Sharma, B.M., Aggarwal, R.K. and Praveen K. (1990). Water retention and nutrient availability in a fly ash amended desert sandy soil : A study in *vitro.* Arid Soil Res. Rehab., 4, 53-58.

Sharma, D.C. and Sharma, C.P. (1996). Chromium uptake and toxicity effects on growth and metabolic activities in wheat, *Triticum aestivum L.* Cv. UP. 2003. Indian J. Exp. Biol., 34, 689-691.

Shinde, B.N., Kochan, S.M. and Patil, A.S. (1995). Use of fly ash for improving physical properties of black soil. J. Maharashtra Agri. Uni., 20:2, 167-169.

Shorrocks, V.M. (1974). Boron Deficiency-Its prevention and cure. Soman-Wherry Press, Norwich, MA.

Shukla K.N. and Mishra, L.C. (1985). Effect of fly ash extract on growth and development of corn and soyabean seedlings. 'Water, Air arid Soil Pollut., 27: 1-2,. 155-167.

Siedlecka, A., Krupa, Z., Samuelsson, G., Oguist, G. and Gardestrom, P. (1997). Primary Carbon metabolism in *Phaseolus vulgaris* plant under Cd/Fe interaction. Plant Physiol. Biochem., 35, 951-957.

Sikka, R. and Kansal, B.D. (1994). Characterization of thermal power plant fly ash for agronomic purposes and to identify pollution hazards. Biores. Technol., 50: 3, 269-273.

Sikka, R. and Kansal, B.D. (1995). Effects of fly ash application on yield and nutrient composition of rice, wheat and on pH and available nutrient status of soil. Biores. Technol., 51:2-3, 199-203.

Sims, J.T., Vasilas, B.L. and Ghodrati, M. (1993). Effect of coal fly ash and co-composted sewage sludge on emergence and early growth of cover crops. Comm. Soil Sci. Plant. Analy., 24:5-6, 503-512.

Sims, J.T., Vasilas, B.L. and Ghodrati, M. (1995). Evaluation of fly ash as a soil amendments for the Atlantic Coastal Plain: II, Soil chemical properties and crop growth. Water, Air and Soil Pollut., 81:3-4, 363-372.

Singh, A.K. Singh, R.B., Sharma A.K., Gouraha, R. and Saggar, S. (1997) Response of fly ash on growth of *Albizia procera* in coal mine spoil and skeletal soil. Environ. Ecol, 15: 3, 585-591.

Singh, D.B., Makkar, K., Verma, S. and Mishra, S.N. (2000). Effects of putrescine, ammonium nitrate, and IAA in ameliorating metal and salinity induced stress in mustard seedlings. Indian J. Plant Physiol., 5, 257-263.

Singh, K.K. and Mishra, L.C. (1988). Reduction in chlorophyll and energy contents in plants as indicators of atmospheric pollution. Photo synthetica, 22:1 129-132.

Singh, N. and Yunus, M. (2000). Environmental Impacts of Fly Ash. In : Iqbal, M., Srivastava, P.S. and Siddiqi, T.O., Environmental Hazards : Plants and People, CBS Publishers and Distributors, New Delhi, 60-79.

Singh, N., Yunus, M. and Ahmad, K.J. (1994). Growth response and element accumulation in *Beta vulgaris* L. rasied in fly ash amended soils. Ecotoxicology, 3:4, 287-298.

Singh, N.B. and Singh, M. (1986). Effect of fly ash application on saline soil and on yield components and uptake of NPK of rice and wheat, at varying fertility levels. Annals of Agr. Res., 7:2, 245-257.

Singh, R.N., Bhumbla, D.K., Keefer, R.F. and Harvath, D.J. (1992). Improving production by altering chemical properties of mineland with industrial wastes. In : Bajwa, M.S., Pasricha, N.S., Sidhu, P.S., Chaudhary, M.R. and Benbi, D.K. (eds.) Proc. Inter. Symp. Nut. Sust. Product. : Vol. 1, 36, 366-380.

Singh, R.P., Saxena, S.K. and Kumari, K. (1991). Effect of fly ash on the growth of tomato plants, nematode population and available nutrients in soil. Inter. J. Trop. Agri., 9:2,. 88-92.

Sohal, G.S. (1995). NTPC Northern Region-Powerful Performance. In: Electrifying success. The Pioneer-A special feature. Nov. 7.

Somani, L.L., Totawat, K.L., Mundra, M.C. and Sharma, R.A. (1994). Amelioration of alkali soils through fly ash. In : Dryland resource and technology. Somani, L.L. and Bansal, K.N. (eds.) 8, 187-193.

Spencer, L.L.S. and Drake, L.D. (1987). Hydrogeology of an alkaline fly ash landfill in Eastern Iowa. Ground water, 25, 519-526.

Srivastava, K., Farooqui, A., Kulshreshta, K. and Ahmad, K.J. (1995). Effect of fly ash amended soil on growth of *Loctuca sotiva* L. J. Environ. Biol., 16:2, 93-96.

Srivastava, R., Kumar, A., Rai, A.M. and Singh, L. (1995). Biological method of fly ash stabilisation through afforestation on ash dump yards near thermal power station, Panki (Kanpur). Indian Forester, 121:2, 81-87.

Steffens, J.C. (1990). The heavy metal binding peptides of plants. Annu. rev. Plant Physiol. Plant Mol. Biol., 41, 553-575.

Stobart, A.K., Girffiths, W.T., Ameen-Bukhari, I and Sherwood, R.P. (1985). The effect of Cd on the biosynthesis of chlorophyll in leaves of barley. Physiol. Plant., 63., 293-298.

Stoewsand, G.S., Anderson, J.L., Weinstein, L.H., Osmeloski, J.F., Gutenmann, W.H. and Lisk, D.J. (1978). Selenium in tissues of rats fed ruta bagas grown on soil covering a coal fly ash landfill. Bull. Environ. Contam. Toxicol., 44, 681-685.

Straughan, I., Elseewi. A.A. Page, A.L. and Hemphill, D.D. Eds. (1978). Mobilization of selected trace elements in residues from coal combustion with special reference to fly ash. In: Trace substance in environmental health, XII, 389-402.

Su, D.C., Zhang, F.S. and Wong, J.W.C. (1997). The growth of *Agropyron elongōtum* in an artificial soil mix with coal fly ash and sewage sludge. Chinese J. Appl. Environ. Biol., 3:3, 230-235.

Su, N.R. (1982). The use of industrial wastes for plant nutrient recycling in Taiwan. In : Extension Bulletin, ASPAC Food and Fertilizer Technology Centre, 178, 1-10.

Sugawe, G.T., Quadri, S.J. and Dhoble,, M.V. (1998). Response of sunflower to the graded levels of fertilizers and fly ash. J. Maharashtra Agri. Uni., 22:3, 318-198.

Summers, R., Clarke, M., Pope, T. and O'Dea, T. (1998). Western Australian fly ash on sandy soils for clover production. Comm. Soil Sci. Plant Analy., 29:17-18, 2757-2767.

Swanson, V.E. (1972). Composition and trace element content of coal and power plant ash. Part 2, Appendix J of southwest Energy study. U.S. Dep. of the interior Geol. Surv., Open File Rep. 61p.

Taets, C. and Rayburn, A.L. (1996). Coal fly ash exposure at agronomic levels does not induced triploidy in maize. Bull. Environ. Contam. Toxicol., 56:5. 690-695.

Taylor, E.M. and Schuman, G.E. (1988). fly ash and lime amendment of acidic coal spoil to aid revegetation, J. Environ. Qual., 17, 120-124.

Terman, G.L. (1978). Solid wastes from coal fired power plants. Use or disposal on agricultural lands. Bull V. 129. T.V.A. Muscle Shoals, Ala.

Tietzeova, E. and Makusova, Z. (1981). The effects of industrial pollutants on the antomical characteristics of the pores of selected gross species. *Sbornik-Vysoka- Skoly-Zemedelske-v-Praze, Fakulta, Agronomicka,* 34 A, 15-43.

Tischner, R. (2000). Nitrate uptake and reduction in higher and lower plants. Plant, cell Environ., 23, 1005-1024.

Tolle, D.A., Arthur, M.F. and Von Voris, P. (1983). Microsm / field comparison of trace elements uptake in crops grown in fly ash amended soils. Sci. Total Environ., 31, 243-261.

Tomar, R.M.. Kaur, I., Neelu and Bhatnagar, A.K. (2000). Effect of enhanced lead in soil on growth and development of *Vigna radiata* (Linn.) Wilczek, Indian J. Plant Physiol., 5, 13-18.

Townsend, W.N. and Gillham, E.W.F. (1975) Pulvensed fuel ash as a medium for plant growth. In : M.J. Chadwick and G.T. Goodman (eds.). The Ecology of Resource Degradation Renewal, Blackwell Scientific, Oxford pp. 287-304.

Townsend, W.N. and Hodgson, D.R. (1973). Edaphological problems associated with deposits of pulverised fuel ash. In : R.J. Hutnik and G. Davis (eds.) Ecology and Reclaimation of Devastated land, Gardon and Breach, New York, 1, 45-56.

Tripathy, A. and Sahu R.K. (1997). Effect of coal fly ash on growth and yield of wheat. J. Environ. Biol., 18:2, 131-135.

Trotta (1940), In: The Wealth of India, Raw materials (1992), Vol.4 (Revised), Publication and Information Directorate, CSIR, New Delhi.

Tung, G., Mc Ilveen, W.D. and Jones R.D. (1995) Synergistic effect of fly ash and SO_2 on development of cucumber leaf injury. Environ. Toxicol. Chem., 14: 10, 1701-1710.

U.S. Salinity Laboratory Staff., (1954). Diagnosis and improvement of saline and alkali soils. Handb. no. 60. U.S. Government Printing Office. Washington, D.C. 160p.

US Environmental Protection Agency, (1988). Wastes from the combustion of coal by electric utility power plants. USEPA Rep. 530-SW-88-002, Washington, DC.

Vaccarino, C., Cimino, G., Tripo do, M.M., Langane, G., Giudice, L. Lo. and Matarese, R. (1983). Leaf and fruit necrosis associated with vanadium rich ash emitted from a power plant burning fossil fuel. Agric. Ecosyst. Environment, 10, 275-283.

Vallace, B.L. and Ulmer, D.D. (1972). Biochemical effects of mercury, Cadmium, and lead Annu. Rev. Biochem., 41, 91-128.

Vallack, H.W. and Chadwick, M.J. (1992). Monitoring air bore dust in a high density coal fired power station region in North Yorkshire. Environ. Pollut., 80, 177-183.

Valtini, G., Vaccari, F., Pera, A., Agnolucci, M., Scatena, S. and Varallo, G. (1999). Evaluation of composted coal fly ash on dynamic of microbial populations and heavy metal uptake, 7:1, 81-90.

Van Hook, R.I. (1979). Potential health and environmental effects of trace elements and radionuclides from increased coal utilisation. Environ., Health Perspect., 33. 227-247.

Vierling, E. (1991). The roles of heat Shock proteins in plants. Annu. Rev. Plant Physiol.-Plant Mol. Biol, 42, 579-620.

Vlamis, J. and Williams, D.E. (1964). Iron and manganese relations in rice and barley. Plant Soil, 20, 221-231.

Vollmer, A.T., Turner, F.B., Straughan, I.R. and Lyons, C.L., (1982). Effect of coal precipitator ash on germination and early growth of desert annuals. Environ. Exp Bot., 22, 409-413.

Wadge, A. and Hutton, M. (1987). The uptake of cadmium, lead and selenium by barley and cabbage grown on soils amended with refuse incinerator fly ash. Plant and soil, 96: 3, 407-412.

Wallace, A. and Wallace, G.A. (1986). Enhancement of the effect of coal fly ash by a poly acrylamide soil conditioner on growth of wheat. Soil. Sci., 141, 387-389.

Wallace, A., Alexander, G.V., Saufi, S.M. and Mueller, R.T. (1980). Micronutrient supplying power of pyrite and fly ash. J. Plant Nutr., 2, 147-153.

Warambhe, P.E., Kene, D.R., Thakare, K.K., Darange, O.G., Thakare, S.K. and Chaphale, S.D. (1992). Nutrient availability and growth performance of cotton receiving fly ash application. J. soils crops, 2:2, 9-11.

Warambhe, P.E., Kene, D.R., Thakare, K.K., Darange, P.G. and Bhoyar, V.S. (1993). Evaluation of physiochemical properties of fly ash of thermal power station Koradi (Nagpur) for its likely use in agriculture. J. soils and crops, 3:1, 75-77.

Weeldreyer, P.D. and Fine, L.O. (1981). Phosphate in waters : II. Plant availability of lignite fly ash extracted forms in greenhouse trials. Water Resour. Bull., 17, 1083-1805.

Williams, A.J., Chandra', K.K., Dugaya, D., Mishra, A.K. and Banerjee, S.K. (1995). Population status of VAM fungi in rehabilitated ash disposal area. Environ. Ecol., 13:2, 384-387.

Wong, J.W.C. (1995). The production of artificial soil mix from coal fly ash and sewage sludge. Environ. Technol., 16:8, 741 - 751.

Zhu, Y., Pilon-Smith, E.A.H., Jouaxin, L, and Terry, N. (1999). Over expression of glutathione in *Brassica juncea* enhances Cadmium tolerance and accumulation. Plant Physiol., 119, 73-79.

Zwick, T.C., Arthur, M.F., Tolle, D.A. and Van Voris, D. (1984). A unique laboratory method for evaluating agroecosystem effect of an industrial waste product. Plant Soil, 77, 395-399.